Quantum Theory and Fre

Henry P. Stapp

Quantum Theory and Free Will

How Mental Intentions Translate into Bodily Actions

 Springer

Henry P. Stapp
Lawrence Berkeley National Laboratory
University of California
Berkeley, CA
USA

ISBN 978-3-319-86370-2   ISBN 978-3-319-58301-3   (eBook)
DOI 10.1007/978-3-319-58301-3

Printed on acid-free paper

This Springer imprint is published by Springer Nature
The registered company is Springer International Publishing AG
The registered company address is: Gewerbestrasse 11, 6330 Cham, Switzerland

Contents

Prologue

A revolution in our scientific understanding of the physical world occurred during the twentieth century. That upheaval revised our idea of science itself, and thrust our conscious thoughts into the dynamical process that determines our physical future. During the preceding two centuries, from the time of Isaac Newton, our conscious experiences had been believed by most scientists to be passive witnesses of a clock-like physical universe consisting primarily of tiny atomic particles and light, that evolves with total disregard of our mental aspects. Our conscious thoughts had, for two hundred years, been exiled from science's understanding of the workings of nature. But during the first quarter of the twentieth century that earlier "classical" theory was found by scientists to be unable to account either for the observed properties of light, or for the plethora of new empirical data pertaining to the dynamical properties of actual atoms such as Hydrogen and Helium. A better theory was needed!

In 1925 Werner Heisenberg, the principal creator of quantum mechanics, concluded from an analysis of the data of atomic physics that the basic precepts of the prevailing classical theory were profoundly wrong, and that the root of the difficulties lay in Newton's ascribing to his conception of atoms certain properties that empirically observed atoms do not possess. The first of these purely fictional ideas is that each atom has at each instant of time, a tiny well-defined location in 3D space. The second is that the evolving physical properties are completely determined by prior physical properties alone, with no input from our conscious thoughts. These latter *"mental"* realities were assumed, in the classical theory, to be completely determined by the *physically described* properties of the associated brains and nervous systems. Hence they do not, in that theory, constitute extra free variables. But in quantum mechanics the evolution of the physically described aspects is not fully determined by physically described properties alone: our conscious experiences enter irreducibly into the dynamics!

The basic principle that guided Heisenberg to the successful new theory was that it should be based on properties that we can choose to measure. These choices on the part of conscious agents are "free choices" from amongst the many possibilities allowed by the theory—of which measurement to actually perform. Here the adjective "free" means that the choices are not determined by purely "physical" laws alone. They are determined in part by irreducible mental aspects of psycho-physical observers. The observer's "free choices" are thus non-physical inputs into the physical dynamics! Our minds are not mere side effects of material physical processes: Our Minds Matter!

Heisenberg's study of the data of atomic physics convinced him that the empirical data of atomic physics cannot be reconciled with the naïve realism of classical mechanics, which considers a person's perception of an external physical property to be a mere causally inert by-product of the observational process of creating a brain/body representation of that property. According to that classical scenario, nature goes to the great length of creating a seemingly new kind of stuff, mental reality, which, however, has no physical function or effect. Such an arrangement seems unnatural.

Quantum mechanics, on the other hand, assigns to mental reality a function not performed by the physical properties, namely the property of providing an avenue for our human *values* to enter into the evolution of psycho-physical reality, and hence make our lives meaningful.

Quantum mechanics accounts with fantastic accuracy for the empirical data both old and new. The core difference between the two theories is that in the earlier classical theory all causal effects in the world of matter are reducible to the action of matter upon matter, whereas in the new theory our conscious intentions and mental efforts play an essential and irreducible causal role in the determination of the evolving material properties of the physically described world. Thus the new theory elevates our acts of conscious observation from causally impotent witnesses of a flow of physical events determined by material processes alone to irreducible mental inputs into the determination of the future of an evolving psycho-physical universe. In this orthodox quantum mechanical understanding of the world our minds matter!

An adequate basic scientific theory of reality must explain <u>all</u> of the regularities of human experience. That totality includes not only data pertaining to the motions of planets and terrestrial objects, and the evidence from atomic physics, but also the evidence concerning the observed effects in our everyday lives of our conscious intentional efforts upon our subsequent bodily behavior. These ubiquitous facts of everyday life exhibit a strong positive correlation between one's conscious intention to produce a desired bodily action—such as the raising one's arm or the moving one's finger—and a follow-up bodily motion of the intended kind. Thus my mental effort to raise my arm is normally quickly followed, if I focus my intention upon it, by the rising of my arm. An appreciation of this correlation between subjective mental intent and subsequent physically described reality is far more important to the normal living of one's life than the periodic motions of some tiny pin-points of light in the night sky. What matters most to us is what we are able to do about our physical future, and how we are able to do it. In this connection, the everyday-experience-based belief in the causal power of a person's mental effort to influence the subsequent physically described reality is rationally buttressed by the fact that contemporary (i.e., quantum) science supports that intuition, rather than diminishing us by claiming, as did classical physics, that the experienced causal effectiveness in the physical world of our mental intentions is "the illusion of conscious will".

A personal belief in the power of one's mental intentions to affect one's physical future is the rational foundation of our lives. Our cognizance of this causal effectiveness of our thoughts underlies our rational effortful engagement with the world, and, consequently, also the structures of our social institutions; of our moral imperatives; of our legal systems; of our notions of Justice; and of our conscious efforts to improve the lives of ourselves and those we care about. In a causally mindless mechanical world of the kind entailed by the materialist precepts of classical physics this power of our minds is denied, and that denial eliminates any possibility of a rationally coherent conception of the meaningfulness of one's life. For how can your life be meaningful if you are naught but a mechanical puppet every action of which was completely fixed by a purely mechanical process pre-determined already at the birth of the universe?

The inclusion of the quantum element of random chance can rescue the meaningfulness of one's life. For these choices are not written in stone, or fixed by mechanical determinism. Hence, by being fundamentally indeterminate, or "random", they can be become biased by *values*, which can thereby influence the course of physical events!

The philosophical difficulties ensuing from Newton's presumption about light and the atomic character of matter are eliminated from orthodox quantum physics by the replacement of the Newtonian classical dynamics by a quantum dynamics that elevates our minds from passive bystanders to active participants in the creation of our psycho-physical future. This radical revision of the role of our minds in the determination of our future arises directly from the elimination, from the material world, of all particles of the kind imagined to exist by Isaac Newton, and their replacement by the quantum idea of "atomic particles". These latter entities are mathematically described elements of a new kind that are intrinsically tied to our conscious experiences. Replacing the purely fictional Newtonian particles by the experience-related atoms of quantum physics converts the classically conceived world that has no rational place for causally efficacious conscious experiences into a quantum world of "potentialities" for certain experiences to occur. It converts a known-to-be-empirically-false materialist conception of the world into a rationally coherent quantum conception of reality in which our causally efficacious minds play an essential role in the determination of our common psycho-physical future.

According to this quantum mechanical understanding of reality, the very same laws that were originally introduced to account for the *empirical findings* in the domain of atomic physics explain also how a person's mental intentions can affect that person's bodily actions in the way that he or she mentally intends. The advance from seventeenth-century materialistic science to twenty-first-century probabilistic quantum physics thus converts our minds from slaves of our brains to causal partners with our brains.

A general recognition of this profound transformation of science's image of man from mechanical automaton to "free" (from material coercion) agent constitutes a contribution of science to today's troubled world that could in the end be far more important than its engineering offerings. For how we use our scientific knowledge depends on our values, and our values depend on our self-image.

The aim of this book is to convey to general reader's, in simple but accurate terms, how realistically interpreted orthodox quantum mechanics works, with emphasis on the impact of this science-based understanding of ourselves on the meaningfulness of our lives.

Chapter 1
The Origins of the Quantum Conception of Man

Every culture has its lore about the origins and nature of the world and its people. Those ideas are often associated with a deity, or deities, and an associated religion. But there arose in western civilization in the seventeenth century, in connection with the ideas of Galileo Galilei and Sir Francis Bacon, the notion of a "scientific" approach to our understanding of the nature of things. Galileo emphasized the importance of doing experiments specifically designed to shed light on particular questions. Thus in order to gain knowledge about how gravity works he measured the acceleration of falling objects of varying weights by dropping them from high places, or by allowing them to roll slowly down inclined planes. Sir Francis Bacon, on the other hand, emphasized that a detailed understanding of the workings of nature would allow us to put nature to work for us: to make her a potent ally in our pursuit of human well being. Thus, whereas our basic beliefs about the nature of things had generally been based on ancient traditions and sacred writings that ratified prayers and acts of worships as the prescribed means of getting nature to help us, the new "scientific" idea was to gain an understanding of the regularities of nature by means of experimental observations, in order to put her thus-discovered orderliness to work for us.

This seismic shift from religious dogma to empirical evidence was the basis of the science that followed. Isaac Newton used it to develop what has become known as classical mechanics, which prevailed as the fundamental scientific theory about the nature of things until the beginning of the twentieth century. But at that point it became clear that nature did not conform to the simple precepts postulated by Isaac Newton. A new scientific theory was needed, and was duly created.

Over the course of the first half of the twentieth century, scientists constructed relativistic quantum field theory (RQFT), which is a hugely successful rational approach that yields validated predictions of high accuracy. The core difference between the newer theory and the older one is that quantum theory is primarily about, and is built directly upon, the empirical structure of our conscious experiences, whereas the classical theory was built on a postulated dynamics of material properties, with the everyday apparent dependence of material properties on our

© Springer International Publishing AG 2017
H.P. Stapp, *Quantum Theory and Free Will*,
DOI 10.1007/978-3-319-58301-3_1

conscious intentions reduced to an asserted dependence upon material properties alone. Thus standard quantum mechanics involves, in an essential way, the causal participation of the minds of us observers, while classical mechanics strictly bans any such effect of mental realities on the world of matter.

I shall begin this narrative with a brief sketch of the more familiar classical physical theory, which is still taught in our schools and some of our colleges without adequate emphasis on its profound differences with its contemporary quantum successor with respect to the causal role of our minds.

The Classical Predecessor to Contemporary Physics

The science-based approach to understanding nature began in earnest with the work of Isaac Newton, who said:

> … it seems probable to me, that God in the Beginning form'd Matter in solid, massy, hard, impenetrable, movable Particles". But the core message of quantum mechanics is that this "solid particle" conception of matter is a figment of Newton's imagination: a pure fiction completely unlike the stuff that constitutes the constituents of the "material world" as it is understood in orthodox quantum mechanics. In that newer theory the mathematically described world in which we find ourselves embedded has the nature of "a set of potentialities for the occurrence of certain kinds of perceptions". And these potentialities behave in many ways more like mental realities than like the solid material particles that Newton described. Moreover, those Newtonian particles were presumed to interact with one another primarily by contact. Yet, according to Newton, they also attract each other by the force of gravity, which acts instantaneously over astronomical distances.

When accused of mysticism because of this assumed instantaneous action at a distance Newton replied: "That one body can act upon another at a distance through a vacuum, without the mediation of anything else…is to me so great an absurdity that I believe no man who has in philosophical matters a competent faculty of thinking can ever fall into it."

Newton obviously rejected as nonsense the idea of an immaterial instantaneous action at a distance. Yet he offered no hypothesis about how the information concerning the location of a source of gravity could be instantaneously conveyed to a faraway system. He justified his mysterious assumption by the fact that it led to an understanding of many known astronomical and terrestrial empirical findings, such as the orbits of planets, the rising of tides, and the falling of apples.

Nothing goes Faster Than Light?

More than two centuries later, Albert Einstein proposed an explanation that made gravity's influence non-instantaneous, and, indeed, transmitted at the speed of light. Einstein's theory demanded, moreover, that *no influence of any kind* could transfer information faster than the speed of light. This condition became a bedrock principle of physics that was generally accepted by scientists. But the challenge of maintaining it in the face of twentieth-century empirical findings (or dealing adequately with its failure) has become the most basic task of the science of our era. Our entire scientific world-view rests upon the completion of this task, which is entangled with our science-based understanding of our own human nature.

Descartes' Dualism

These issues concerning the basic nature of things were brought into focus, before Newton was born, by the writings of the great French philosopher and mathematician René Descartes. He argued that what exists is divided into two different kinds of things: 'things that occupy locations in three-dimensional space at instants of time', and 'entities that think'.

This Cartesian duality set the stage for the developments of science that followed. It allowed the conceived reality to be divided, actually, into _three_ different kinds of things: _material properties_, _mental realities,_ and _thinking entities_. _Material properties_ are features of things that are built out of particles and their associated energy-carrying fields, and that are fixed by the properties of these component particles and fields. _Mental realities_ include your thoughts, ideas, and feelings. A _thinking entity_ is an entity that is _experiencing mental realities_.

An example of a possible Cartesian _material property_ is the location of a tiny Newtonian-type particle whose center is located at each instant of time at a point in 3-D space, with the rest of it lying nearby. Two examples of _mental_ realities are your feeling of pain when you touch a hot stove, and your experience of the color "red" when looking at a ripe tomato. An example of a _thinking entity_ is the "You" that is now experiencing the reading of this book: it thinks your thoughts, knows your ideas, and feels your feelings. It is the "I" of Descartes' famous "I think, therefore I am."

Descartes recognized that the mental events occurring in a person's stream of conscious experiences are associated with the material processes occurring in that person's brain. But he maintained that these mental realities are fundamentally different in kind from the corresponding material activities in the brain. This difference is the famous, or infamous, Cartesian distinction between mind and body, or mind and brain.

Classical Determinism

Isaac Newton, building on Descartes' ideas, focused his attention on the material aspects. He formulated mathematical "laws of motion" that account in a detailed way for the motions of the planets in the solar system, for the orbit of the moon around earth, for rising tides and falling apples, and for a host of other observed features of the "material" universe. This account makes no mention of any influence of mental realities upon material properties, and is called "classical mechanics" or "classical physics".

By virtue of these laws, applied universally to all material things, living or dead, a classical Newtonian-type universe is "deterministic". This means that the entire history of that universe is fixed for all time, once the initial conditions and the mathematical laws of motion are specified. The aspects of the material universe that are not fixed by the general laws are thus limited to the selection of the initial conditions and the choice of the (assumed time-invariant) laws of motion. Specifying these two inputs then determines every material event that will ever occur. No matter-based feature is left to chance, or to the will of either Man or Nature. This early-science-based way of trying to understanding reality in terms of matter alone, with no essential input from a mental realm, is called "materialism", or sometimes "physicalism".

Philosophical Torment

Philosophers have been tormented for centuries by this seeming verdict of science that reduces human beings to mechanical automata. Our rational thoughts and moral sentiments were rendered incapable of deflecting, in any way, our bodily actions from the path ordained at the birth of the universe by the purely machine-like material aspects of nature. That conception of reality destroys the rational foundations of moral philosophy: How can you be responsible for your actions if they were completely determined before you were born, and, indeed, at the birth of the universe?

This torment is not confined to moral philosophers. The great nineteenth-century physicist John Tyndall touched upon it when he wrote:

> We can trace the development of a nervous system and correlate it with the parallel phenomena of sensation and thought. We see with undoubting certainty that they go hand in hand. But we try to soar in a vacuum the moment we seek to comprehend the connection between them... (The Belfast Address, 1874).

The core difficulty here is that mental realities, which certainly do exist, have no rational place within the framework of 17th/19th century science. They are logically disconnected appendages that are added on, ad hoc, simply because we know that they exist. But their effects on what happens in the material world are, according to classical mechanics, the same as if they do not exist. The reason we seem to be 'soaring in a vacuum', as Tyndall bemoans, comes from the materialistic viewpoint of classical mechanics. That way of thinking, in order to be complete, must permit the existence of the thoughts we actually experience. Yet it provides absolutely no logical foundation, or even tiny toehold, for any rational understanding of how human consciousness or feelings can arise from the logical foundation provided by the materialistic precepts of classical mechanics.

The Copenhagen Shift to a Pragmatic Stance

During the first quarter of the twentieth century, a series of experiments were performed that probed the properties of matter at the level of its atomic constituents. The results were incompatible not merely with the fine details of classical mechanics, but with its basic tenets as well.

Responding to this catastrophic breakdown of classical mechanics, scientists created, during the first half of the twentieth century, a new theory called "quantum mechanics". It is based on concepts profoundly different from those of classical physics, yet yields extremely accurate predictions about the outcomes of all reliably replicable experiments, both old and new. It leads also to a revised understanding of our own human nature that is radically different from the effectively mindless mechanical conception entailed by the materialistic principles of classical mechanics.

The original version of quantum mechanics is called "The Copenhagen Interpretation" because it was hammered out in intense discussions centered at Niels Bohr's institute in that city. In order to dodge various philosophical difficulties, quantum theory was originally offered not as a "theory of reality", but rather as a "pragmatic set of rules". These rules were designed to allow physicists to make reliable statistical predictions about what observers will experience in response to their various *contemplated* alternative *possible probing actions of observation or measurement*.

Virtues of Realism

But the new theory can also be interpreted "realistically", or "ontologically", as "an understanding of reality itself". A realistic interpretation is, in fact, needed if one seeks to extract from science any deep insight into the nature of the universe and of our human selves within it. The thesis expounded in this book is that von Neumann's orthodox formulation of quantum mechanics, elucidated where needed by the ideas of Heisenberg, Dirac, Wheeler, and the mathematician, logician, and philosopher Alfred North Whitehead, and updated to the relativistic form developed by Tomonaga and Schwinger, can be regarded as a theory of reality that is sufficiently detailed and accurate to deal with the issues of the general nature of our mental aspects, and of the causal connection of our conscious minds to the material world in which our brains and bodies are embedded.

A Condition on the Scope of a Science-Based Theory of Reality

An adequate scientific theory of reality ought to accommodate <u>all</u> the regularities of human experience. This includes not only the results of experiments pertaining to astronomical, terrestrial, and atomic physics, but also to the experiences of normal everyday life. These ubiquitous subjective data reveal a strong positive correlation between a person's felt mental intention to perform a simple bodily action, such raising an arm or a finger, and a subsequent perception of the intended bodily action!

Empirical data of this kind constitute the rational foundation of our active meaningful lives, for they effectively instruct us how, by making appropriate mental efforts, to influence our bodily actions in mentally intended ways. A theory of reality that fails to provide a rationally coherent account not only of astronomical, terrestrial, and atomic data, but of also this directly experienced mind-body relationship, is fundamentally deficient. Such deficient theories include materialistic classical mechanics, which claims that everything real is created by the interaction of matter with itself, but then fails to explain how these purely material processes generate our conscious perceptions and our causally efficacious mental efforts. Similarly inadequate is any non-standard materialistic version of quantum theory that does not account for our subjective experiences, and the capacity of mental effort to influence in desired ways the behavior of our bodies!

The standard "orthodox" quantum mechanics can, by virtue of its mathematical structure, and the words used to describe it, be naturally interpreted realistically, and when thus-interpreted it brings our mental aspects into the dynamics as elemental realities that are causally linked to matter via specified "laws of nature". I call this interpretation "Realistically Interpreted Orthodox Quantum Mechanics". It evades the logically impossible task of explaining how felt mental properties can be constructed out of mechanical material properties alone, by postulating the elemental existence of both mind and matter, and then describing in rational mathematical terms how they interact with each other.

Von Neumann's "Orthodox" Formulation of Quantum Mechanics

The "standard" quantum theory, against which all others are compared, is von Neumann's "orthodox" formulation of Copenhagen Quantum Mechanics, or, more specifically, the updated version, called "Relativistic Quantum Field Theory", abbreviated as "RQFT". It is this relativistic "orthodox" version of quantum theory that is propounded in this book. As will be presently explained, this theory is about both: (1), the dynamical interaction of matter with itself that accounts for the 'unobserved' behavior of material substances; and (2), the interaction between mind and matter that constitutes the highly nontrivial 'process of observation'.

In quantum mechanics the mind-matter interaction is mathematically very different from the matter-matter interaction. And it is different in a mathematical way that entails that the former can never be reduced to the latter. The difference in these two dynamical processes is directly connected to Heisenberg's seminal 1925 discovery, which quickly led to the creation of quantum mechanics. This new theory gives detailed explanations of the plethora of twentieth century data of atomic physics that had resisted all attempted explanations via the materialist precepts of classical physics. Heisenberg's discovery was that the process of observation—whereby an observer comes to consciously know the numerical value of a material property of an observed system—cannot be understood within the framework of materialist classical mechanics. A non-classical process is needed. This process does not construct mind out of matter, or reduce mind to matter. Instead, it explains, in mathematical terms, how a person's immaterial conscious mind interacts with that person's material brain.

An immaterial mind lies beyond the ken of a materialistic approach, and the mathematics that describes the process of conscious observation is not reducible to the mathematics that describes the process of the unobserved evolution of matter.

The eminent Hungarian-American mathematician and logician John von Neumann cast the ideas of Copenhagen quantum mechanics into a rationally coherent and mathematically rigorous form that is widely used by mathematical physicists, and also by others who require mathematical and logical precision. Nobel Laureate Eugene Wigner labeled Von Neumann's formulation "Orthodox Quantum Mechanics". The label "Orthodox" is appropriate, in the sense that many, and perhaps all, mathematical physicists take it to be the logically and mathematically precise formulation of the Copenhagen ideas.

Von Neumann approached these mind-related issues by considering what amounts to a tower of good measuring devices where each device associates, one-to-one, each input to a corresponding output, and the output of each device is the input to the device above it. On the top of this tower lies an observer's conscious "ego" that can both receive perceptual inputs and instigate probing actions by means of its interactions with its associated brain.

About the entry of consciousness into the dynamics, von Neumann says:

> First, it is inherently entirely correct the measurement or the related process of subjective perception is a new entity relative to the physical environment and is not reducible to the latter. Indeed, subjective perception leads us into the intellectual life of the individual, which is extra-observational by its very nature [vN p. 418].

This first quote emphasizes that, within von Neumann's "orthodox" representation of quantum mechanics, the process of subjective perception is not reducible to the process that governs the interaction of matter with itself. Our subjective conscious perceptions are, as Descartes had declared, neither equivalent to, nor reducible to, the behavior of matter. I take this irreducibility of mind to the behavior of matter to be, on the basis of this quote—and everything else said in von Neumann's book—a core feature of realistically interpreted "orthodox" quantum mechanics.

The second quote is:

> ...it must be possible so to describe the extra-physical process of subjective perception as if it were in reality in the physical world—i.e., by assigning to its parts equivalent real parts in the objective word in ordinary space. [vN p. 419].

I take these "equivalent real parts" to be, primarily, the neural (or brain) correlates of our conscious perceptions.

The third quote is:

> Now quantum mechanics describes the events which occur in the observed portion of the world, so long as they do not interact with the observing portion, with the aid of Process 2, but as soon as such an interaction occurs, i.e., a measurement, it requires the application of Process 1. [vN p. 420]

This third quote introduces the two very different processes: Process 1 and Process 2. Process 2 is the quantum analog of the dynamical process of classical physics. Like its classical counterpart, Process 2 involves only the material aspects of nature, and is deterministic. It is also "unitary", which means, essentially, that its action merely shuffles information around without losing any of it. This Process 2 depends in no way on the mental aspects of nature. But the material/physical state of the universe, upon which Process 2 acts, contains the neural correlates of our perceptions that were introduced in the second quote.

Process 1 is the process that generates perceptions. Each Process-1 action is associated with a particular conscious observer. It has a mathematical form that is very different from that of Process 2. Process 1 is not "unitary" but is, instead, "projective": it is associated with the subjective occurrence of a perception coupled to the instantaneous elimination from the material universe of all aspects that are incompatible with the occurrence of that perception. Thus this process has two phases. The first phase selects a possible next subjective perception on the part of the observer. This 'possible/potential' next perception defines a corresponding brain correlate, which has, according to the theory, a certain statistical weight. The second phase of Process 1 then reduces the material universe to two parts, one that definitely contains this brain correlate and the other that definitely does not contain this brain correlate, and it "actualizes" either one part or the other. This choice made by nature between the two parts accords with a certain statistical rule known as the "Born Rule". This Born-Rule choice is the (unique) place where "an element of chance" enters into the quantum dynamics. The preceding choice of a possible next perception reflects the history and the felt values of the observer, and is identified with what the observer feels is his or her personal subjective choice of what physical property of the observed system to probe or inquire about. No element of chance is ascribed to this choice made by an observer of a particular possible probing action.

The two phases of Process 1 are manifestations of the differing points of view of Heisenberg and Dirac, cited by Bohr, in which Heisenberg emphasized the free choice on the part of the experimenter of which probing experiment to perform, while Dirac emphasized the choice of the part of nature regarding which outcome occurs.

The whole process resembles, as emphasized by Wheeler, the game of twenty questions, in which a succession of Yes/No questions is posed, with each eliciting 'Yes' or a 'No' response. Von Neumann cast into rigorous mathematical form the key ideas of the founders, insofar as they strayed from the official "pragmatic" path, and tried—as scientist rightfully do—to understand what is really going on.

The fact that the generation of a conscious perception involves a dynamical process that is structurally and mathematical extremely different from the deterministic matter-driven process that governs the unobserved evolution of reality is the basic difference between materialistic classical mechanics and its quantum successor. The classical theory presumes that all aspects of nature can be explained purely in terms of the action of matter upon matter. But the quantum world differs in a fundamental way from that core precept of materialism. This huge structural difference in the real (i.e., quantum) world between the matter-matter interaction and the matter-mind interaction makes manifest the extreme naiveté of trying to comprehend the connection between mind and matter within a materialistic framework.

In the quantum world the observing processes of acquiring empirical knowledge must disturb, or perhaps even bring into existence, the values that we observe. By virtue of Heisenberg's discovery, the process of our acquiring knowledge about the material aspects of nature cannot merely reveal already existing values. The process of our acquiring knowledge injects our mental aspects in an essential way into the process that determines "what we will find if we look".

This non-materialistic action injects the mind of the observer as a causal agent into realistically interpreted orthodox quantum mechanics. It gives our minds an essential dynamical role to play, and hence a natural and rational reason for them to exist.

In this game of "twenty questions" the 'Yes' answer is the occurrence some particular perception, say "P". So the question must, in principle, be whether the upcoming experience will be "P"? The "question" is thus a (non-verbalize) inquiry of the form "Will my upcoming perception be "P" ", where "P" is a felt/experienced representation of a particular possible next perception. This query is instantly followed by "nature's" response, 'Yes' or 'No'. This two-phased process allows our human conscious choices to enter causally into the evolution of the matter-based aspects of the world, rather than being helpless witnesses of a flow of events completely determined by the material aspects of nature alone.

Our probing actions and their observed outcomes are described in terms of "potential" and "actual" perceptions, respectively. According to the orthodox theory, these perceptions are described in the language (conceptual structure) of classical physics. In orthodox quantum theory the disparate perceptual and associated material properties are causally tied together by the quantum dynamical laws that govern their mutual interaction: mind is dynamically tied to matter, but it is neither made of matter; nor dynamically pre-determined by matter.

In summary, the orthodox quantum understanding of the evolving world rests upon a specified quantum "process of evolution" of a *psycho-physical* universe. This process consists of two very different sub-processes. Von Neumann calls them Process 1 and Process 2. Process 2 is the quantum analog of the classical process of evolution of material systems. It produces evolution in accordance with the famous Schrödinger equation. This Process 2, by itself, generates a completely specified and pre-determined continuous temporal morphing of the material properties of the universe into a continuous "quantum smear" of classically describable possibilities or potentialities. Like its classical counterpart, this process depends in no way on any mental aspect of nature. This Process-2 matter-generated evolution generates, however, not just one single world of the kind that we actually perceive, but rather a 'continuum' of possible perceivable worlds. Consequently, some other process is logically required in order to extract from this Process-2 generation of a continuum of "potentialities", a choice of what actually happens in the material world, besides this Process-2 generation of ever-growing sets of perceptual possibilities. This other process is called "Process 1". This numbering, which might seem odd, reflects the fact that the very first action had to choose and actualize some particular state of reality, not just shuffle around information that was already present.

Our personal conscious thoughts enter the quantum dynamics via these abrupt Process-1 actions. Our human minds instigate (probing) actions upon the quantum atomic-particle-based material world that is evolving in accordance with Process 2. Each such Process-1 intervention is, in line with the ideas of Heisenberg, Dirac, Wheeler, and Whitehead, resolved by a two-phased action. The first phase is a probing action, which poses a Yes/No query about an observer's upcoming perceptual experience. The choices of these probing actions are, according to the precepts of quantum mechanics, not fully determined by the prior material properties of the atom-based world. They originate in association with the observer's mind, which von Neumann calls the observer's "ego".

These choices are "free" in the very specific sense that they are not determined by the prior quantum mechanical state of the universe. They are therefore called "free choices". This injection of an element of freedom (from material coercion) constitutes a major departure of quantum mechanics from classical mechanics, and from the general philosophical stance of "materialism", which demands that the evolution of matter be fully determined by material properties alone.

The second phase of Process 1 is an immediate 'Yes' or 'No' response on the part of what British physicist Paul Dirac, a key founder of quantum mechanics, called "nature". A positive 'Yes' response adds the perception P, which was specified by the observer's query, to the observer's stream of conscious perceptual experiences. It also instantaneously reduces (i.e., collapses) the global quantum state of the universe to the part of its immediately prior form that is compatible with that positive response. A negative response leaves the observer's stream of consciousness unaffected (no perception occurs). But it reduces the global quantum state of the universe to a form compatible with that negative response. Nature responds sequentially to the probing yes/no questions posed by the various observers.

But how is the needed connection established between a person's mental choices of probing actions and the intended bodily response?

It can be assumed that the observer's ego creates, by trial-and-error learning, beginning in the womb, a mapping of the perceived response to each of various mentally instigated probing actions. Thus the ego's knowledge of which effort tends to lead to which perceptual feedback can be learned: It need not be innate.

[In mathematical terms, the quantum mechanical state of the universe, ρ, is first reduced by the observer's probing action to a sum of two terms $(P \rho P)$, and $(P' \rho P')$, where this P is a "projection operator" (P times P equals P) and P' is $(1-P)$. Nature "actualizes" one or the other term in statistical concordance with the "Born Rule", which asserts that the probability that state $(P \rho P)$ will be actualized is Trace $(P \rho P)$ divided by Trace ρ, where, for any operator or matrix M, Trace M mean the sum of the diagonal elements of any matrix representation of M.

I have included this parenthetical mathematical remark merely to assure readers that the "ordinary words" that I have been using are not mere verbal fluff. They have definite mathematical meanings, which lead to predictions that are accurate, in one highly non-trivial case, to the width of a human hair, compared to the distance to the moon, and that encompass in general a vast realm of pertinent empirical phenomena. Quantum theory thus warrants serious consideration by any reader truly interested in the basic nature of things. Any adequate proposed alternative to the orthodox interpretation needs to produce a theory of the mind-brain connection that is as good, or better, than the orthodox theory described here. For our experiences are the only things we know, and hence their empirical structure needs to be explained by any basic physical theory that can be deemed satisfactory.]

The Action of Mind on Matter

The entry of the abrupt Process-1 actions into the continuous Process-2 evolution entails that the mind of an observer is no longer a helpless witness to a mechanically predetermined course of material events. The Process-1 actions convert the observer's ego into an actor on the world stage. Each probing action, initiated by an ego, influences—by means of nature's response to that action—the macroscopic behavior of the atomic-particle-based material universe. Thus our minds become endowed, by means of the quantum mechanical dynamical rules, with the power to influence the macroscopic properties of matter, without themselves being totally predetermined by material properties alone!

This empowerment of our psychological aspects is a fundamental feature of standard quantum mechanics. This change in the basic ontological structure eliminates, by means of an advance in science, the absurdity of a consciousness that can do nothing but delude us into believing an outright lie—that our minds are causally inert—which, if believed, would surely be detrimental to the ambitions of anyone who accepts as veridical the findings of science. For an acceptance of the belief in the total physical impotence of one's conscious efforts would seriously undermine the mental resolve needed to overcome the obstacles that often stand in the way of our efforts to create what we judge to be a better world.

The empowering message of quantum mechanics is that the empirical data of everyday life, and also our intuitions, are generally veridical, not delusional; and hence that our mental resolves can often help bring causally to pass the bodily actions that we mentally intend. The role of our minds is to help us, not to deceive us, as the materialist philosophy must effectively maintain.

Appearances are Deceiving: Classical Appearances Versus Quantum Realities

According to the quantum rules, described above, each true perception is an experience of certain macroscopic features of a material (i.e., atomic-particle-based) universe. But a basic precept of quantum mechanics asserts that our perceptions are described in the language of classical physics. This means that, according to quantum theory, each perception P is described in terms of macroscopic properties of systems built basically out of the "solid, massy, hard, impenetrable, movable particles" of Isaac Newton. That world of perceivable macroscopic properties can be called the world of "appearances". On the other hand, quantum mechanics also describes the macroscopic properties of macroscopic systems built out of the atomic particles of atomic physics. From the point of view of realistically interpreted orthodox quantum mechanics, the underlying physically described reality is the quantum state (the so-called density matrix) of the universe, and it is built out of quantum mechanically represented <u>atomic particles</u>, and their associated physical fields, whereas the appearances (perceptions) are built out of <u>Newtonian particles</u>.

But this means that in quantum mechanics both the 'physical reality' and the 'appearances' are represented mathematically. This 'mathematical duality' provides the foundation for a greater role of mathematical and logical rigor than was possible in the classical-physics-based materialistic approach. There the underlying 'physical reality' is deemed to be built out of the fictitious Newtonian particles (instead of the contemporary-science-based atomic particles), and the appearances are described in psychological terms. Hence the quantum mechanical approach to the mind-brain problem is structurally and mathematically very different from the classical/materialistic approach, and—because both our perceptions and the underlying material causal structure are described mathematically—it provides for greater mathematical and logical rigor than classical physics allows.

These two physical theories, classical and quantal, are contradictory. Yet orthodox quantum mechanics combines them to produce a rationally coherent understanding of the connection between mind and brain. This quantum approach constitutes a way of comprehending that connection that is far more reasonable than what is attainable within the materialistic framework, which is fatally flawed by the omission of our causal minds from the theory of the mind-brain connection: "It's like Hamlet without the Prince of Denmark."

The oft-heard claim that "quantum mechanics is not relevant to the mind-brain problem because quantum theory is only about tiny things", is absolutely contrary to the basic quantum principles. Being 'big' does not tend to make a quantum system truly classical! Quantum mechanics is explicitly designed to cover 'big' systems, but by becoming 'big' a quantum system does not become classical!

Indeed, the fact that quantum mechanics is explicitly designed to cover big things is important to the solution of the mind-brain problem. For the quantum mechanical dynamics leads to the evolution of the brain, via Process 2, into a mixture of many different brain states that correspond to many different potential experiences, and hence to the need for the added Process 1 that selects for consideration some perceivable small part of the existing mixture, which nature will then promptly either actualize or reject.

This proliferation in the brain of representations of many different alternative possible immediate courses of action is assured by the structure of ion channels. Ion channels are large brain molecules, each having a small tube (a channel) through which ions of a particular kind—say calcium ions—can flow single file, under specific brain conditions, into the interior of a neuron, where they tend to cause that neuron to release, in due course, a "vesicle" of a neuro-transmitter molecules into the gap that separates that neuron from a neighboring neuron. The narrowness of the ion tube ensures that the ion that enters the interior of the neuron has a large uncertainty in its direction of motion. Hence each ion channel in the brain is a source of dynamical uncertainly in the Process-2-generated evolution of the quantum state of the brain. The resulting macroscopic state of the brain will thus tend to evolve into a quantum "mixture" of many different classically describable brain states, each with a different perceptual correlate, between which the mind-dependent quantum Process 1 is free to choose. Thus the pertinent-for-us essence of quantum mechanics is the causal dynamical linkage that QM specifies between our conscious thoughts and our atomic-particle-based brains.

The quantum mechanically entailed causal effects of our mental intentions upon our material brains is in complete harmony with our normal intuition, which is based on our lifetimes of first-hand empirical evidence. Our minds are promoted by quantum dynamics from the absurd role of impotent witnesses of events they cannot affect to causally effective instigators of intended bodily actions. Our minds thus have a natural reason to exist, which is to help us to achieve what we value, not to deceive us into believing we are something we are not!

The conclusion here, and in what follows, is that the realistically interpreted orthodox quantum conception of reality provides not only dynamical explanations of all well-established ordinary empirical data, but, automatically, also the foundation of a rationally coherent dynamical understanding of how our conscious minds can affect our material brains, and hence our material bodies, in ways concordant with both our conscious intentions and the empirical data of everyday life. Those ubiquitous first-hand data, which seem to confirm the causal power of our mental intentions, need not be interpreted as "illusions" or "delusions", as the Newtonian-particle-based materialistic physics appears to demand. Likewise, the problem of the seeming incompatibility of "free will" and "determinism" is resolved by noting that the QM law of evolution incorporates the inputs from our "free" (not-materially-coerced) choices into the causal dynamics! Hence there is natural causal evolution, and thus no causal gap or incompatibility that needs to be explained.

The reader might be encouraged to take von Neumann's formulation of quantum mechanics seriously by considering the words of the distinguished Nobel Laureate Hans Bethe, who said "I have sometimes wondered whether a brain like von Neumann's does not indicate a species superior to that of man." Another expression of the same idea was a (joking) suggestion that von Neumann was actually an outer-space alien who had trained himself to perfectly imitate a human being in every way.

Potentia

The central theme of the realistically interpreted orthodox quantum mechanics (RIOQM) being expounded in this book is that the quantum state (i.e. density matrix) of the universe is not merely a useful pragmatic tool, as proposed in the 'epistic' Copenhagen conception of quantum mechanics, but is also a representation of essential aspects of reality itself.

A basic question is then "What is the ontological character of this aspect of reality?" The answer, in concordance with the ideas of Heisenberg and others, is that it has the character of set of "Aristotelian Potentialities". That is, it is a collection of potentialities, or tendencies, or proclivities, or dispositions, or probabilities for each one of a collection of alternative possible actual mental events to occur, each in conjunction with new updated set of potentialities. This transformation must be actualized by a process. In RIOQM one such process consists of a two-phased Process-1 probing action of the kind described in orthodox von Neumann quantum mechanics. There might conceivably be other actualization processes in the fullness of nature, but they are neither specified nor taken into consideration in this theory, which is focused on the details of the connection between our causally efficacious conscious minds and our quantum mechanically described brains.

Summary

This first chapter has provided a quick overview of Realistically Interpreted Orthodox Quantum Mechanics. The next nine chapters weave into the narrative more details of the various key features of the orthodox causal connection of our minds to our brains. Chapter 11 goes a step further, ontologically, by arguing that the detailed behavioral properties of the various parts of the orthodox structure suggest that the mental and material aspects of reality are lodged in an overarching nonlocal reality that is fundamentally mind-like in character: the mental and material aspects of the quantum dualism tend to merge, upon detailed analysis, into an underlying mental monism that includes mathematically described properties conceived to be embedded in a 4D space-time continuum. In the end it is indeed true that "all is one", and that that "unity" encompasses both our mental and material aspects.

That idea, that the underlying reality is fundamentally mind-like, has been advanced often before, on the basis of all sorts of reasons. But that conclusion is often thought to involve going beyond science, and, indeed, going against science. Here that conclusion arises from science-based considerations alone.

Chapter 2
Waves, Particles, and Minds

Particles and Waves

Classical mechanics developed during the nineteenth century—due principally to the work of James Clerk Maxwell—into a form that involved two different kinds of physical stuff: "particles" and "waves". Electrons are the prime example of particles, whereas "light", in the form of the electromagnetic field, is the prime example of a wave. Particles are tiny, highly-localized structures, each with a center that, at each instant of time, is situated at one precise point in three-dimensional space, with the rest of the particle lying nearby. A wave, on the other hand, tends to spread out over a large region in space, and to exhibit "interference patterns" due to the cancellations or reinforcements of moving crests and troughs.

Particles and waves have, therefore, contradictory structures: particles always stay tiny, whereas waves tend to spread out. Thus Planck's discovery in 1900 that light, which had seemed to be a wave, had a corpuscular nature came as quite a shock. Light of a given frequency appeared to be emitted in chunks, each carrying a quantity of energy that is directly proportional to the frequency of the light wave, with a universal proportionality factor called Planck's constant. Albert Einstein won the Nobel Prize for his explanation, five years later, of the photo-electric effect. Empirically, a metallic surface radiated by light of a definite frequency emits electrons with energies equal—after a correction for the energy needed to get the electron out of the metal—to the energy of the incoming quantum of light, now understood to be localized like a particle.

The concepts of classical physics were unable to cope either with this wave-particle-duality problem, or with a large number of other problems concerning the properties of atoms. A new way of understanding nature was needed, and was created.

© Springer International Publishing AG 2017
H.P. Stapp, *Quantum Theory and Free Will*,
DOI 10.1007/978-3-319-58301-3_2

Science and Philosophy

These problems of wave-particle duality and atomic structure appear to be completely physical in character. But the founders of quantum mechanics were men of profound philosophical bent. Niels Bohr's father was an eminent physiologist familiar with the writings of William James, and Wolfgang Pauli was the godson of the philosopher Ernst Mach. Werner Heisenberg, whose father was also a professor, was greatly influenced by the views of Bohr and Pauli. All three were strongly influenced by the view of Albert Einstein that science rests in the end on empirical findings, and that our physical theories are basically human inventions that help us deal with the world known to us only via our conscious observations and experiences. Bohr, concurring, announced at the start of his 1934 book, *Atomic Theory and the Description of Nature,* that "In physics...our problem consists in the coordination of our experiences of the external world." A few pages later (p. 18) he writes:

> In our description of nature the purpose is not to disclose the real essence of phenomena, but only to track down as far as possible relations between the multifold aspects of our experience.

In line with this viewpoint, the founders of quantum theory officially presented their theory not as what would normally be called a description of an existing and evolving material reality, as was done in classical mechanics. Their theory was offered, rather, as a tool that scientists have invented for making testable and useful predictions about future experiences on the basis of knowledge gleaned from prior experiences.

That official position was a secure one from which Bohr could defend the theory against Einstein's objections. It was useful also for keeping students on a productive track of learning how to use the theory in practical applications, and preventing them from spending (wasting?) their time pondering philosophical issues about which even the founders did not fully agree. Heisenberg and Pauli both devoted much time and effort trying to understand the nature of the reality lying behind the pragmatic rules. And von Neumann speaks in his discussion of the measuring process about the connection of the "intellectual inner life of the individual" to the circumstances "which actually exist in nature." He seems very clearly to be talking about an underlying reality, not merely a pragmatic tool.

The fate of classical mechanics provides a stark warning of the danger of taking initial success as tantamount to victory in the search for truth. Accordingly, the impressive empirical successes of standard (Copenhagen-Orthodox) quantum mechanics have failed to convince all physicists of the real need to bring into the dynamical laws any experiential quality that is not fully specified by the material and space-time structure of the universe. Alternatives to standard quantum mechanics have thus been proposed that are essentially in line with the precepts of materialism, which exclude from the dynamics all immaterial elements. But the theme of this book is that von Neumann's (orthodox) formulation of quantum mechanics, as elucidated herein, has, by virtue of the rational coherence of its mathematical, empirical, and philosophical components, the qualifications that warrant its being regarded as an adequate putative theory of reality itself. A "sine qua non" of an 'adequate' theory of reality is that it provide a rationally coherent understanding of the relationship between our conscious experiences and the associated processes occurring in our brains.

The Realistically Interpreted Orthodox Quantum Mechanics described here violates the demand of materialism that our conscious experiences have no causal power beyond what can be explained by the causal properties of matter alone—where 'matter' consists of things described in geometrical terms, and built out of geometrical structures like Newtonian particles and their associated energy-carrying fields.

This quantum mechanical world is basically a *psycho-physical* structure in which the causal effects of the disparate mental and atomic-particle-based elements are woven together by means of von Neumann's carefully formulated quantum dynamical laws. Those laws entail that a person's material actions can be influenced in specified ways by *his or her mental aspects in ways that are not fixed by the evolving material aspects of the universe alone. This understanding of standard (Copenhagen-Von Neumann) quantum mechanics is thus fundamentally non-materialistic: our mental aspects enter into the evolution of matter in ways not reducible to effects of matter alone. It is an understanding that is based on the words and concepts of the founders—particularly Heisenberg's and Bohr's reference to the "free choices" of probing actions on the part of experimenter-observers, and Dirac's choice of response on the part of "nature", all rigorously expressed in the mathematics and words of John von Neumann.*

This insertion of fundamentally mental causes into our basic physical theory generates a gross violation of what had, for two hundred years, been widely regarded as a key feature of a 'scientific' theory of reality; a feature considered to identify a proposed theory as science, as opposed to non-science. Indeed, the materialist demand of strict exclusion from the material world of all effects of mental causes is still regarded as a scientific imperative by many researchers, who consequently endeavor to explain the seemingly mind-related behavior of a person's body, whilst stoutly denying the possibility of any actual causal effect of that person's mind upon his or her bodily behavior.

But how did this radical break with materialism ever come about? How and why did the band of highly reputable physicists that created quantum mechanics suddenly, in 1925, feel entitled to make this huge break with the then-highly-honored classical materialistic tradition? The answer is to be found in:

Heisenberg's Seminal 1925 Discovery

The common idea of quantum mechanics in the minds of many non-physicists centers on Bohr's renowned model of the atom. According to that model, atoms are like miniature classical solar systems in which the circling electrons tend to stay on favored orbits, but make occasional "jumps" from one such orbit to another, with an associated emission or absorption of a photon. That model is an essentially classical physical system, with some added "quantum" conditions that there exist these favored orbits whose locations are related to the mysterious quantum constant discovered in 1900 by Max Planck.

Bohr's model dates from 1913, and hence was twelve years shy of the 1925 creation of quantum mechanics. While that 1913 model certainly does bring an important quantum element into the dynamics, it is seriously deficient as a characterization of the essential difference between classical mechanics and its quantum successor. It is ironic that this Bohr model of orbiting electrons is often offered as an example of the quantum nature of things, when, actually, the creation of quantum mechanics, triggered by Heisenberg's 1925 work, was precisely a rejection of the ideas of the 1913 quasi-classical Bohr model, with its definite trajectories of orbiting electrons, and lack of all reference to "our knowledge".

The key differences between standard Copenhagen/Orthodox quantum mechanics and its classical predecessor are, first, that the classical notion of particles as tiny objects moving on trajectories is replaced by the quantum notion of atomic particles represented by waves; second, that in the new theory these particles do not have well-defined trajectories; and third, that the needed abrupt collapses of the quantum states of systems are instigated by mental aspects of nature, not by purely mechanical/material aspect of nature acting alone. Thus our conscious experiences are, according to the new orthodox view, not causally inert bystanders, as in classical mechanics, but play an essential causal role in the determination of the objective psycho-physical future. These differences underscore the radically new ideas that emerged from Heisenberg's 1925 discovery, and that are mathematically embodied in realistically construed standard (Copenhagen/Orthodox/RQFT) quantum mechanics.

The principle of the "causal closure of physical" is, as mentioned earlier, sometimes regarded as part of a definition of science: a discriminating property that sets science apart from non-science. But science is perhaps better characterized, following the leads of Galileo and Bacon, by our essential use of probing actions intended to test hypotheses, and thereby allow us acquire knowledge about the material world; coupled with our practical applications of the knowledge that we thereby acquire.

Bohr's 1913 model does not bring into the dynamics any clear indication of a failure of the core precepts of materialistic classical physics. It merely adds some quantum conditions. And that model seemed to be putting physics onto a promising track. But then how and why did this radical triad of ideas (the representation of an atomic particle by a wave; the omission of particle trajectories; and the essential

location & momentum

incorporation into the dynamics of the non-materialistic process of our acquiring knowledge) suddenly become accepted in 1925 by the founders of quantum mechanics as core precepts of their new physical theory? How did those completely alien and subversive ideas gain traction in a scientific environment so intrinsically hostile to it?

This abrupt 1925 turnabout was instigated by the persisting failures of the semi-classical attempts to account for the accumulating data of atomic physics, coupled with a profound discovery made in 1925 by Werner Heisenberg. He had come to believe that something was profoundly wrong with the (essentially classical) ideas of the 1913 Bohr model, and that the needed new theory should be built on properties that are actually known to exist—by virtue of our capacity to become cognizant of their numerical values by performing appropriate measuring procedures. These considerations directed Heisenberg's attention to the empirical processes of acquiring knowledge. While studying, theoretically, the processes of measuring, respectively, the 'location' and the 'momentum' of an atomic particle, say an electron, Heisenberg found that if the 'location' was measured first, and the 'momentum' second, then the product of the two outcomes differs from the product obtained when the two properties are measured in the reverse order. And the difference between these two products is essentially the famous constant that Planck discovered in 1900. Consequently, this completely unexpected connection between the outcomes of the two observation procedures must be connected to the quantum character of reality. And it entails that the process of acquiring knowledge about material properties cannot generally leave those properties undisturbed! For, if the process of acquiring knowledge allowed the observer simply to become aware of fixed pre-existing values then the two products of the outcomes could not remain differing by the fixed Planck's constant *in the limit in which the times of the two measurements tend to become equal. Heisenberg discovered that our actions of acquiring knowledge must disturb the observed system* in detailed ways *that are intricately tied to Planck's constant!*

That discovery quickly led Heisenberg, Born, and Jordan to a radically new theory based on the idea that, in keeping with certain prevailing philosophical ideas, the core subject matter of a satisfactory theory of the nature of things should be 'the evolving structure of our empirical knowledge of the world'—not 'the evolving structure of an imagined material world built primarily upon Newton's "solid, massy, hard, impenetrable, moveable particles'. Those particles can reasonably be viewed as pure fictions that happen to be useful in certain macroscopic contexts, but that fail to work in situations involving our acquiring of knowledge about the structure and behavior of atomic particles, particularly those contained in the neural/brain correlates of our perceptions.

The notion that the material world is built (principally) out of these Newtonian particles is, from the standard view of QM, a useful fictional creation of Isaac Newton. There exists no empirical evidence for their actual existence. Accordingly, the core subject matter of the new theory is taken to be something we do know, namely the structure of our evolving knowledge of the material world. This knowledge is asserted to be generated by the specified "objective mind-brain process of acquiring subjective knowledge". This process of observation is, according to the new theory, instigated in part—just as we innately feel it is—by the observer's mental intent and conscious effort, which thereby causally affect the observed material world. Orthodox QM spells out in great—although not complete–detail of how this mind-brain connection works.

Using measuring devices to acquire knowledge about matter dates from antiquity. And telescopes and microscopes were important in the development of classical mechanics. But in quantum mechanics Heisenberg's discovery entails that, in principle, these two processes of measurement—of 'location' and 'momentum'—cannot individually always leave the measured system just as it was, and with definite values of these two properties. For, if they did, then the product of the outcomes of these two knowledge-acquiring operations would have to be independent of the temporal ordering of these two procedures, in the limit in which they became simultaneous.

Thus Heisenberg's 1925 discovery entails that the increases in our knowledge of the properties of matter, which we acquire by performing measurements, cannot in general leave the state of the measured matter unchanged, and with definite values of these two properties. The probing processes that allow us to gain knowledge about properties of matter must 'in principle' sometimes 'disturb' those properties by finite (non-zero) amounts specified by Planck's constant. But in classical mechanics this difference can in principle be smaller than what quantum reality demands! Thus, in order to accommodate Heisenberg's finding, about the mind-brain connection we must, as a matter of principle, abandon classical mechanics, and, more generally, the philosophy of materialism!

The problem facing the founders was not merely to acknowledge the failure of the simple idea that we trivially acquire knowledge of the material world by simply mentally grasping directly the material facts, as was effectively assumed in classical mechanics. It is obvious that the fact that we can learn about the motions of the tiny pinpoints of light that correspond to planets, without appreciably affecting their motions, does not automatically carry over to the motions of the points that correspond to the locations of the electrons or atoms in our brains. The needed quantum theory had to account for the fact that the process of acquiring knowledge about the properties of the material world had to disturb the material structure in precisely the quantitative way needed to account for Heisenberg's findings! Thus a major revision in our understanding of the mind-matter connection lies at the heart of quantum mechanics.

To expect, under these conditions, to understand the mind-brain connection within the materialistic classical framework is truly an "Astonishing Hypothesis"—as was recognized by Francis Crick, who nevertheless espoused it, and called for a classical-physics-based neuroscience. That recommendation has dominated subsequent neuroscience, and produced a plethora of data, but, unsurprisingly, no understanding of how our mental consciousness is connected to our material brains. This book is about the non-astonishing orthodox QM claim that the mind-matter connection is a quantum effect.

In the light of Heisenberg's discovery, the founders of quantum mechanics were emboldened to let go of classical mechanics, which effectively sets Plank's constant to zero, in conflict with nature, and, instead, build a rationally coherent alternative to classical mechanics that incorporates into its foundational structure Heisenberg's discovery pertaining to the general non-trivial effects of the process of acquiring subjective knowledge about the objective state of the material world, and that moreover permits precise predictions about the observed structure of human knowledge. Within this quantum framework a person's acquired knowledge of material properties is not a faithful representation of the pre-probing properties of the observed system, but is, instead, an output of a dynamical probing processes initiated by the observing person. The observer's un-coerced-by-matter choices of what to observe affect the temporal evolution of the material aspects of nature.

One therefore cannot exclude the effects of the processes of our acquiring knowledge from of an adequate basic physical theory. That effect is both limiting and liberating: it limits, via the uncertainty principle, what we can know, but expands, via the entailed power of our minds, the possibilities for what we can do!

The orthodox quantum framework is, therefore, not just an arbitrary construct conjured up out of thin air by the founders, *and justified merely by its eventual success in accounting for the behavior of matter*. The driving endeavor of the founders was to create a rationally coherent conceptual structure that accommodates and explains—and is able to make useful predictions about—the structure of our conscious experiences. Our experiences thereby become the basic veridical realities of the theory, not misleading delusions.

Heisenberg's 1925 discovery was that the process of acquiring knowledge about the material world is very nontrivial! It is not a mere grasping of preexisting realities, but a highly structured action upon those realities. That unexpected result elevates the science-based conception of ourselves from passive observers to active agents. That reversal is the underlying core message of quantum mechanics! In the oft-cited words of Niels Bohr: "In the drama of existence we are ourselves both actors and spectators."

Standard quantum theory is thus a psycho-physical (or perhaps an episto-material) theory of the interaction of the evolving material aspects of nature with our evolving knowledge of those aspects. The theory, with its detailed agreements with observed (hence macroscopic) data, emerged, basically, from Heisenberg's guiding principle, which restricts what the theory 'postulates to exist" to properties of a kind that we can, via our observations, 'know exist'. His principle was to build on an empirically secure foundation, instead of empirically unsupported guesses.

The close agreement of the resulting theory with the normal objective empirical data is certainly a bottom-line success. But standard quantum theory describes, via Process 1, also the dynamical connection between a person's mentally instigated actions and that person's consequent mental perceptions of material responses to those actions. Any putative alternative "non-standard quantum theory" that fails to provide a rational theory of these more subjective aspects of the mind/brain connection is fundamentally deficient, compared to the standard quantum mechanics.

It was the assumed possibility for an ideal observer to know in principle, simultaneously, both the 'location' and the 'momentum' of every particle in the universe (and eventually the analogous properties of the fields) that allowed Laplace to deduce from the materialist principles of classical mechanics the "determinism" of the material world, and hence, within the framework of classical mechanics, the impossibility of a causal intervention of anything not fully characterized by its material properties. But that whole notion of the "causal closure of the physical" fails in a world where the mind-dependent quantum dynamical rules prevail.

We do not directly perceive atomic particles. We perceive only 'big' (macroscopic) systems that are built out of combinations of large numbers of atomic particles (and their associated physical fields). Quantum mechanics has well-defined rules for combining many atomic particles together to make big objects and systems, and to represent in mathematical language the purely mechanical (Process-2) aspect of the evolution of those macroscopic systems.

A 'big' physical object, although perceived in classically describable terms, is not causally governed by the laws of classical physics. It must be treated as a conglomeration of its atomic (quantum) constituents in order to account for its physical properties such as rigidity and electrical conductivity. Yet if it is treated as a conglomeration of its atomic quantum mechanical constituents evolving in accordance with Process 2 alone, then it will not have in general, and most specifically when it is a measuring device, a classically describable location and shape. Process 2 generates a quantum state (i.e., density matrix) that represents a sum (called a "mixture") of a 'continuum' of potential/possible worlds of the type that we can actuality perceive or experience, but does not specify which element (or set of elements) in this continuum will be "actualized" if someone looks.

This "mixture" of potentialities is sometimes called a "smear" of potentialities. Thus the quantum mechanical state of the macroscopic "pointer" on a measuring device is, by virtue of the process-2 evolution, "smeared out" over a continuous collection of potential locations along the dial. But that whole smear is not what is perceived if someone looks. It is the mind-dependent Process 1, not the mind-independent Process 2, that resolves the question of what our actual experiences are.

Process-2 evolution includes the interaction of the system of interest with the surrounding environment, but that "environmental decoherence" effect falls far short of specifying what an observer will experience/perceive if he looks! It is Process 1, not mere environmental decoherence, that provides that needed result.

As already described in Chap. 1, this Process 1 first selects, from the Process-2-generated continuum of potentialities, a particular perception that 'might' occur. Then 'nature' chooses, subject to the statistical Born Rule, either to accept the possibility selected by the observer, and then actualize the global consequences of that acceptance, or actualize the global consequences of rejecting the observer's proposal.

[The above description decomposes the standard vN description of an event that can involve, all at once, a large set of possibilities, into an ordered sequence of possibilities, each involving a single Yes/No question, as in the game of twenty questions. Thus the whole large set questions can be considered to be posed, one-by-one, with no passage of physical time, until a 'Yes' response eventually appears. This easily graspable formulation, proposed by Wheeler, is equivalent to the standard one, and more easily converted to the relativistic version demanded by RQFT. That latter version of the theory requires that a particular 3D "global instant now" be defined in association with each of nature's Yes-or-No responses, and that the associated global collapse be instituted along that 3D surface, which divides 4D space-time into an associated "past" and an associated "future". This will be discussed later.]

By means of the two processes, Process 1 and 2, the standard (Copenhagen-Von Neumann) approach elevates our inner mental selves, our egos, from passive spectators to active agents. From this orthodox quantum mechanical perspective, the basic difficulty with putative materialistic versions of quantum mechanics that leave our human mental choices out of the dynamics, is that they leave the theory burdened with (1): our useless conscious processes; and (2), a quantum mechanically evolving world with no means for selecting, from the Process-2-generated quantum smears of possibilities, what our actual perceptions will be. Moreover, the denial of the causal potency of our mental efforts is blatantly contradicted, empirically, by the ubiquitous experiences of everyday life. The materialists' claim that this experiential basis of our lives is an "illusion" rings hollow when the theory that makes this claim is found to be false, and is replaced by a hugely successful theory in which the ubiquitous daily experiences of the causal power of our mental intentions in the world of matter is rationally explained.

The Standard Copenhagen-Von Neumann Approach

The aforementioned 'smearing' difficulty is resolved in the standard quantum approach by bringing into the dynamics something beyond the Schrödinger equation, namely the probing actions of observing agents. The probing query might be, "Will my upcoming experience be that of the pointer on the measuring device lying between 5 and 5.01 on the dial?" A 'Yes' response on the part of nature consists of nature's delivering to the observer the query-defined possible experience, and reducing the quantum state of the entire universe to the part of its prior self compatible with that 'Yes' response. A 'No' answer will result in a corresponding reduction, but no immediate experiential feedback. This omission leaves room for another query to be posed with no passage of physical time. Thus millions of 'No's' can be produced by Nature with (little or) no passage of measured physical time.

The primary reality assumption in the realistically interpreted orthodox quantum field theory that I am describing is that the evolving quantum state (i.e., density matrix) of the universe is an element of reality. The behavior of this quantum state is concordant with the idea that it represents, as Heisenberg and the philosopher Alfred North Whitehead have emphasized, a collection of (Aristotelian) "potentialities for future experiences". This quantum potentiality normally evolves according to the definite Process 2, but, in order to become an 'actuality', a 'potentiality' must be 'actualized' by some other process, and the future is thus considered to be "open". In contrast, a future 'classical possibility' is mechanically predetermined to 'happen' or 'not happen' already at the birth of the universe, thereby precluding any possibility that our mental intentions and efforts can make any difference in what happens to our physical bodies.

In von Neumann's formulation, the purely atomic-physics-based dynamical process (Process 2) does not fail because a system is 'big'. It fails because the atom-based aspects of the dynamics are *only part of the causal story*. The causal, deterministic unitary Process 2 is disrupted by non-unitary Process-1 perceptual observations, which therefore have causal effects upon the physical/material world that are not caused by the purely matter-based Process 2. Thus materialism fails!

That is, the purely matter-based Process-2 evolution fails when that evolution comes into causal contact with the material correlates of our subjective experiences, which are the neural (or brain) correlates of our subjective experiences of probing and perceiving. No other failure of Process 2 is mentioned!

Von Neumann spends a lot of time and effort reducing the quantum mathematics to properties of so-called projection operators. These can be directly related to experiments that have just two alternative possible results, Yes or No, which can be associated with whether or not an observer perceives a specified response or fails to perceive such a response to his probing action. This association allows well-defined connections to be formed between von Neumann's mathematics and observer perceptions. If the answer is 'Yes' then the specified perception occurs. If the answer is 'No' then no perception occurs, for no perception can be all the perceptions other than the specified one.

This rule allows many immediate 'No' responses to be delivered by nature before the one 'Yes' in a multiple-choice question.

The purely mechanical atom-based Process-2 evolution fails when a measuring process is performed, due to the over-riding character of the Process-1 action.

Orthodox quantum mechanics is thus basically a description of this causal dynamical interaction between our conscious minds, which carry our perceptions, and our material atom-based brains, which contain the brain correlates of our probing actions and the responding perceptions.

The earlier classical mechanics is constitutionally unable to accommodate the twentieth-century empirical data. But the most elemental and ubiquitous source of empirical data is one's own daily experiencing of the ability of one's mental effort to influence one's bodily action. Who has not witnessed the intense struggle of the newborn infant to learn, by trial and error, which mental effort produces which perceived bodily response? To classify this first-hand empirical data as an "illusion" in order to salvage a theory that is known to be fundamentally false, and false in a way that is essentially an incorrect understanding of the connection between our conscious experiences and their brain counterparts, is neither rational nor scientific.

The quantum resuscitation of the causal power of our thoughts overturns the absurd classical notion that nature has endowed us with conscious minds whose only power and function is to delude us into believing that it is helping us to create a future that advances our felt values, while in actuality that future was predetermined 15 billon years ago.

Realistically interpreted orthodox quantum theory thus provides us with a non-materialistic science-based understanding of our own intrinsic nature. It is a theory that accounts *with spectacular accuracy* for the structure of the empirical facts about the external world discovered by atomic physicist during the twentieth century. Many competent physicists struggled unsuccessfully for a quarter of a century to comprehend those facts in every imaginable way concordant with the materialistic word view, until Heisenberg, in 1925, lifting that restriction, but clinging to the principle that the new theory should be built upon "observables", and hence in some way upon us observers, broke the log-jam in such a decisive way that Pauli, Born, Jordan and others immediately jumped on board. Einstein, already in 1928, nominated Heisenberg, Born, and Jordan for the Nobel Prize, which was awarded to Heisenberg in 1932. The strangle-hold of materialism was broken simply by the need to accommodate the empirical data of atomic physics, but the ontological ramifications went far deeper, into the issue of our own human nature and the power of our thoughts to influence our psycho-physical future.

Chapter 3
The Measuring Process

Minds, Brains, and Meaning

The final chapter of von Neumann's book is entitled "The Measuring Process". But the real topic is "us", and our acquisition of knowledge. The core message of quantum mechanics, in the words of Niels Bohr, is: "In the drama of existence we are ourselves both actors and spectators." It is our influence on our acts of acquiring knowledge that allows us actors to transform a quantum world of potentialities into actualities that are expressions of our values. It is these consequences of our probing actions that give meaning to our lives.

Utility of Quantum Smearing

The feature of quantum mechanics that converts us from puppets to protagonists is "quantum smearing". This property stems directly from Heisenberg's uncertainty principle. It allows our minds to be more than mere cogs in a giant machine, or helpless witnesses to events they cannot affect. Quantum smearing gives us important things to do, and the dynamical laws of orthodox quantum mechanics endow our minds with the power to do them.

Einstein's First Example

Einstein [5] offers a helpful example of quantum smearing. Suppose a single radio-active nucleus is placed in a sphere and surrounded by a decay-detecting device that, when activated by the decay of the nucleus, sends a signal to a mechanism that causes a pen to make a blip (a spike) on a moving scroll.

© Springer International Publishing AG 2017
H.P. Stapp, *Quantum Theory and Free Will*,
DOI 10.1007/978-3-319-58301-3_3

Application of the quantum analog of the classical laws of motion, namely the Process-2 Schrödinger equation, causes the physical blip to occur not just at one single place on the moving scroll, but at a continuum of locations, each corresponding to a possible time at which the nucleus might decay. If we then follow, via Process 2, the flow of information about the quantum state of the blip-containing scroll into the brain of the observer, we find that the brain will contain, for each blip location in the quantum smear, the neural correlate of the perception of the location of that blip. But the observer will actually perceive the blip to lie at a single small portion of the smear of possibilities. Thus Process 2 cannot be the full story.

According to orthodox quantum theory, the dynamical partner of Process 2, namely Process 1, inserts into the evolution of the quantum states an action (a collapse) that is instigated and partially specified by the mind of the observer. Thus, the observer's mind, or 'ego', is actively involved in reducing the quantum smear of potential perceptions to the single perception that an observer actually experiences. Without these 'smears' there would be nothing for us to do: everything would be pre-ordained, as in classical mechanics. Moreover, the important concept of 'probability' enters into the quantum dynamics precisely at the Process-1 action of reducing the smear of potential perceptions to the one perception that is actually experienced. Without the prior 'smear' of possibilities there would be no place, or role, for quantum probabilities.

We are faced at this point with a deep problem that, in one way or another, has challenged philosophers since the beginning of philosophy, namely the problem of the nature of the connection of our conscious thoughts and perceptions to the material world. Now, however, we are armed with a highly developed mathematical structure that is focused precisely on this issue. As Dirac remarked in the preface to his 1930 book:

> This state of affairs is very satisfactory from a philosophical point of view as implying an increasing recognition of the part played by the observer in himself introducing the regularities that appear in his observations...

Von Neumann starts his discussion of the measuring process by emphasizing that we generally inform ourselves about the physical world by means of perceptions of properties of systems located at some finite distance from our bodies. The pertinent perceptual information about the system being examined is transported to our brains by a sequence, or chain, of intermediate physical systems. Von Neumann illustrates this point by describing a situation in which the information about the temperature of liquid in a container is transferred to the observer's brain by a path that goes first to a thermometer—a column of mercury—then to a beam of light, then to the retina, then to the optical nerve, etc., and finally to a brain structure that is the neural (or brain) correlate of the observer's knowledge of the temperature of the liquid. Each physical system in the chain can, under certain specified conditions (described below) be regarded as part of a good measuring device that transmits the key perceptual information from input system to output system without significant loss.

Quantum mechanics focuses primarily on relationships between our various perceptions, and it is the faithful mapping of the perceptual structure to equivalent forms along the chain that is of immediate interest here. An adopted "principle of psycho-physical parallelism" asserts that this faithful perceptual chain is accompanied by a parallel quantum mechanically described chain that carries the associated probabilistic information.

Von Neumann developed a detailed quantum model of a good measuring device. If one considers a pair of simple quantum system as a single new quantum system, and places it in a so-called "pure state", then the one-to-one mapping between the states of the two sub-systems—input and output—that is needed for a good measuring device will prevail. Thus, in our example, the information about the blips on the scroll will be carried, step-by-step, to associated features of the brain of the observer. Most importantly, the quantum mechanically described chain that parallels the perceptual chain will ensure that the statistical weights (probabilities) associated with the different possible perceptions will be preserved. That is, there is a probability preserving mapping of the perceived aspects of the external perceived scene to corresponding aspects of the brain of the observer.

The Process-1 action that occurs in the observer's brain, can thus be mapped (backward-in-time and outward in space) via this chain of good measurements to a faithful image of the Process-1 action occurring in the observer's brain to any one of the measuring systems along the chain, and ultimately out to the blips on the moving scroll.

The Movable Heisenberg Cut

As already explained, quantum mechanics, unlike its classical forerunner, adopts the view that science is properly about "our knowledge" of the underlying matter-based reality, not directly about that material reality itself. And Heisenberg's key 1925 discovery was that in the quantum universe these two parts of nature differ in very important ways. According to quantum mechanics, the mathematical (Hilbert-space) structure of the underlying atom-based reality is very different from the mathematical (4D space-time) structure of our conscious perceptions of that reality. Hence, a person's mind cannot simply perceptually grasp, directly, the structure of the underlying quantum reality, because the quantum mechanical structure of an observer's brain is incommensurate with the classically describable structure of that person's perceptions. Some mind-brain linking process is needed!

In order to deal with this central problem in a rationally coherent and practically useful way Heisenberg proposed that we conceptually divide reality into two separate parts: (1), an atomically constituted and quantum mechanically described observed system; and (2), a perceptually constituted and classically described observing system. Von Neumann's proof shows that we can, in each observation, shift the placement of the "Heisenberg Cut" between these two parts to any chosen position—along the tower of linked measuring systems that connect the perceived system to its correlate in the observer's brain—without altering the statistical (Born-Rule) weights of the various alternative possible outcomes of that observation. One can consider the mind-matter transition to occur at any link in the chain of "devices" without altering the predictions of the theory.

The proof rests firmly on the postulate that the basic causal dynamics is specified by the orthodox quantum dynamics, even though the conscious perceptions are experienced and described in terms of the classical-mechanical concepts.

Von Neumann describes the situation thusly:

> Now quantum mechanics describes events which occur in the observed portion of the world, so long as they do not interact with the observing portion, with the aid of Process 2, but as soon as such an interaction occurs, i.e., a measurement, it requires an application of Process 1.

This means that in the sub-systems in the tower that lie "external", or "below", the Heisenberg cut one can use, for the individual perceivable possible outcomes, their quantum mechanical descriptions, which specify their individual statistical weights. But "above" the Heisenberg cut, including the brain itself, one can use the perceptual description.

Our capacity, as theorists, to choose "which description to use where" is essential, because we do not know the detailed quantum counterparts of our various possible perceptions, yet need to know their statistical weights in order to be able to make statistical predictions about the various alternative possible experiential outcomes of our alternative possible probing actions.

This statistical information is available to us theorists precisely because we are able to use the (experiential) perceptual description in the brain-side (or upper-side) of the Heisenberg cut, but the quantum description—which carries the statistical weights of the perceptual possibilities—on the (external/objective) lower side of the cut, which contains the perceived physical system. The two descriptions are two aspects of one possible response to a quantum probing action.

But what are those predictions? What determines, in our example, the theoretical probabilities associated with the various alternative possible perceivable locations of the blip on the moving scroll?

The probability that the blip will appear in a specified small region on the scroll is determined by the decay rate of the radioactive nucleus. The (exponential) decay process causes different statistical weights to be assigned to decays occurring during different possible time intervals: the probability for the blip to appear in any small time interval will fall off with time, as the source decays. Because the scroll is moving, the probability will fall off also with a shift of the location of the blip on the scroll. So the probabilities of different possible perceptions (of the blip location) are unambiguously specified by the locations of the blips 'out there' on the scroll, where they can later be perceived by the observer.

Thus the probabilities of the occurrences of the various alternative possible perceptions need not be computed in terms of the probabilities of the neural correlates of those perceptions. They can be computed, as just explained, in terms of properties of features of the perceived scene. The quantum dynamics of the chain of good measurements will then transfer this statistical information into the brain of the observer, where it can influence nature's (later) response to the observer's probing question.

It is, of course, extremely important that scientists be able, as in our example, to deduce the predicted probabilities of possible Process-1 generated perceptions from the associated properties of the perceived world, rather than from the detailed properties of the neural correlates of our perceptions. That is because we lack the capacity either to theoretically know, or to experimentally measure, in sufficient detail, the neural correlates of our perceptions. The spectacular successes of quantum mechanics are not tied to a comparable understanding of the neural correlates of our perceptions. The successes of quantum mechanics are based on the statistical relationships between properties of our perceptions, not on a detailed understanding of the brain correlates of those perceptions! This whole scheme works because of the proof by von Neumann of the movability of the Heisenberg Cut.

I have discussed this proof in some detail here both because of its central importance in orthodox QM, but also in order to dispel any possible misunderstanding of that proof that might suggest—because of its use of the classical/perceptual description on the upper/brain side of the cut—that macroscopic brain dynamics can be described classically. Von Neumann's proof implies no such thing! Macroscopic brain dynamics is quantum brain dynamics: Von Neumann's use of classicality in that proof is not a license to treat the dynamics of our macroscopic brains classically.

Chapter 4
Quantum Neuroscience

"The overwhelming question in neurobiology today is the relationship between the mind and the brain." These are the words of Francis Crick [2]. In the same venue, famed neuroscientist Antonio Damasio [3] writes that the mind-brain question "towers above all others in the life sciences".

Given this recognized major importance of the mind-brain problem, you might think that the most up-to-date, powerful, and appropriate scientific theories would be brought to bear upon it. But just the opposite is true! Most neuro-scientific studies of this problem are based on the precepts of nineteenth century classical physics, which are known to be fundamentally false. Most neuroscientists follow the recommendation of DNA co-discoverer Francis Crick, and steadfastly pursue what philosopher of science Sir Karl Popper called "Promissory Materialism". The "promise" is the long-unfulfilled promise that rigid adherence to the precepts of materialistic classical physics, which exclude minds from the dynamics, will lead to a solution of the problem of the connection between our minds and our brains.

Unsurprisingly, this fundamentally incorrect classical physics has been completely unable to explain how your brain could be or produce your mind. The occasional references to quantum mechanics that one finds in neuroscience are concerned mainly with small-scale behavior at the molecular level, rather than with the core issue, which is the connection of the total behavior of a person's brain to that person's conscious thoughts.

What is the rational basis of the policy of replacing, as the basis of neuroscience, our current hugely successful fundamental theory, quantum mechanics—which is explicitly about the dynamics of the mind-brain connection—by the failed nineteenth century classical mechanics that excludes our minds from the dynamics?

The underlying reason why most scientists, and most Philosophers who are interested in the mind-brain problem, adhere to this failed classical approach seems to be their common belief that the "big" (macroscopic) features of brain dynamics can be adequately described in terms of the concepts of classical physics—that only tiny atomic-sized things need to be described in terms of the concepts and laws of quantum physics. But standard (Copenhagen-von Neumann) quantum mechanics

© Springer International Publishing AG 2017
H.P. Stapp, *Quantum Theory and Free Will*,
DOI 10.1007/978-3-319-58301-3_4

says no such thing! Thus, insofar as we scientists admit that the contemporary pertinent science should be used "in principle", or at least tried, we should seek to explain the empirical data of neuroscience and psychology in terms of the processes 1 and 2 described by von Neumann, rather than presume, because classical mechanics worked for two hundred years, that the notion that our immaterial minds can affect our material bodies must be some sort of illusion. The examined data should include empirical findings pertaining to the connection between perceived scenes and their neural correlates, and encompass also the every-day empirical connections between a person's mental intentions pertaining to his or her bodily actions and the associated observed bodily actions.

In short, if one is interested in the mind-brain connection then the brain must, according to contemporary basic science, be described quantum mechanically! The associated classical description pertains to our perceptions, not the related brain dynamics. So any attempt to understand the mind-brain connection that takes the classical description to be describing our material brains, rather than our conscious perceptions, conflicts with basic contemporary physics, which specifies that von Neumann's dynamical rules—involving Processes 1 and 2—should be used to describe the mind-brain connection. I take that approach to define "quantum neuroscience".

What's the Matter with Matter?

The quantum mechanical description of the atom-based "material" reality is a "bottom-up" description. It is erected upon the quantum mechanical representations of the underlying atomic particles and fields. The quantum rules specify not only how to mathematically describe such elementary quantum systems, but also how to describe systems built out of combinations of many such quantum systems. The dynamical rules specify also how these larger systems behave.

Our brains are made of the same kind of atoms that are studied in atomic physics. The quantum description can thus be extended "all the way up", so as to include the entire world atomic-particle-built stuff, including the body and brain of every observer. If science is rational, then it would be unacceptable for science to simply baldly assert that the laws suddenly fail when things get 'big'. Why do they fail? How do they fail? How big is 'big'? And how about the properties of big things that depend upon the properties of their atomic constituents?

Problems with Materialism

There are four main problems with taking matter, as understood in materialistic relativistic classical mechanics, to be the sole ontological foundation of reality. The first problem is that this classical concept provides no rational roots for the existence of our conscious thoughts, ideas, and feelings. There is nothing in the classical conception of nature that provides any hint of any rational need for, or rational foundation of, the existence of our known-to-exist mental aspects, which thus need to be added, 'ad hoc', to the materialistic ontology—which makes it non-materialistic. The second problem is Heisenberg's 1925 discovery that the process of acquiring knowledge of the material reality necessarily alters that reality in a way that depends on Planck's constant: the connection between the material reality and our knowledge of it cannot be the simple action of direct knowing; some assumption pertaining "knowing" or "experienced knowledge" needs to be added. Quantum mechanics with its Processes 1 and 2, describes a complex process of gathering knowledge that allows our mental intentions to influence our bodily actions in intended ways not allowed by the precepts of classical mechanics. The third problem is the conclusion established in Appendix 1. It demands that faster-than-light (FTL) transfers of information cannot be banned in a context completely described in terms of macroscopic classical concepts. But such FTL transfers are banned in relativistic classical physics. Thus this FTL conclusion *falsifies the belief that the connection between mind and brain can be understood within the framework of relativistic classical mechanics. This FTL problem will be discussed in detail later. The fourth problem is that the classical (Newtonian) building blocks of reality are "solid", whereas the atomic building blocks, being the basis of mere "potentials for psycho-physical events" are evanescent. The Newtonian particles are solid components of a supposed enduring physical reality itself, whereas their atomic quantum replacements are components of mere fleeting potentialities for what the future might be. The two conceived realities are fundamentally different and incompatible, not merely different views of the same reality.*

Einstein's Second Example

A second example of a quantum collapse mentioned by Einstein concerns a mouse and the moon. Suppose there had been, since the birth of the universe, nothing that interrupted its evolution in accordance with the (Process-2) Schrödinger equation. Then the quantum state of the moon would be "smeared out" over the entire night sky, until the first observer, say a mouse, looks. Indeed, the mouse itself would be a "smear" of copies of itself, and the city it inhabits would be a "smear" of all possible cities (etc.), and similarly for the earth, for the solar system, for the galaxies, etc.

In order to cope with the gross mismatch between actual human experience and the matter-based aspects of the temporal evolution of the quantum world, the creators of quantum theory went far beyond the precepts of classical physics. They introduced into the quantum dynamics essential "acts of observation", each associated with a psycho-physical part of reality identified as an *observing agent*.

Each such act is the initiation by the agent of a particular probing action. This probing action "puts to nature" a particular question. As in the game of twenty questions, each question is of the kind that is answered by either a "Yes", or a "No". Multiple-choice situations can be accommodated by decomposing each answer "No" into two parts: a "Sub-Yes" and a "Sub-No" etc. This creates a multiple-choice scenario, with many possible "Yes's" followed by one "No".

Thus we have a question-and-answer scenario, where the questions are 'freely chosen' by a mental 'free agent, and answered and physically actualized by a global mind-like (God-like) agent. But what is the character of this global process that chooses, and then physically actualizes, the *answers*?

Nature's Random Choices

According to quantum mechanics, the *answer* to the question is determined by "a choice on the part of Nature". The answer "Yes" is revealed to the probing agent by the entry into his or her stream of consciousness of a perception that he or she was asking about. For example, if the question is, "Is that fire engine 'red'?" then Nature's positive answer will be revealed to the agent by an experience of 'redness' added onto the previously existing perceptual form of the fire engine. In the version described here, negative answers are not experienced by the probing agent. This allows for many negative responses to occur between any two positive responses. In any case, "experienced reality" is created by dialogs between localized probing agents and a global aspect of reality called "Nature". The probing and answering processes have certain characteristic properties that will be discussed presently.

The agent's "choice" of question is, as already mentioned, not determined by any known rule, and is thus called "free". But Nature's choice of response is subject to certain definite conditions. According to the orthodox theory, Nature's "choice" is "random". This means that in each individual instance, specified by an actual choice of a question, the answer is indeterminate: it is not determined by the dynamical rules of the theory. However, the theory does impose statistical conditions on long strings of instances. For example, the predicted ratio of answers "Yes" to answers "No" in a long string of "replications" is, according to the theory, determined by mathematical properties of the physical state of the system being probed. But which answer actual appears in an individual actual instance is not determined by anything specified in the theory.

The infamous quantum element of "randomness" enters quantum mechanics precisely in this way—and only in this way—through statistical conditions on Nature's choices. The mathematically determined evolution of the physical state via the fixed Schrödinger equation certainly plays a very important role in quantum theory. But the connection of this mathematics to our experiences depends heavily also on these two choices, the first of which is "free", and the second of which is "random".

Relativistic Version (RQFT)

The original versions of Quantum Mechanics were non-relativistic. The relativistic generalization, Relativistic Quantum Field Theory (RQFT), was created in the late 1940's—independently—first by S. Tomonaga and later by J. Schwinger. A key new feature is best described by comparing RQFT with the non-relativistic version. In the non-relativistic version, each measurement event (consisting of a posed question and a response by Nature) was assumed to *occur globally over all space "at an instant of time"*.

In the relativistic version of RQFT, each measurement event is again mathematically implemented by a "collapse/reduction" of the quantum state of the universe that occurs at a single "instant". That "instant", however, is a non-flat 3D surface $[\sum(n)]$ that covers all of 3D space, but with different spatial 3D points allowed to lie at different times—subject to the condition that no point on the non-flat "instant" can be reached from any other one without ever traveling at the speed of light, or faster, or backward in time.

The basic process of the creation of the evolving physical state of the universe is then "forward-in-time", in the sense that each global "instant" is related to the unique preceding one by, at *some* spatial point, (x, y, z), lying 'later in time' than the preceding instant, but at *no* spatial point (x', y', z') lying 'earlier in time' than the time at that point (x', y', z') in the 3D surface that constitutes of the preceding instant.

Between any two successive instants, the quantum state evolves via a generalization of the Schrödinger equation, which is the quantum mechanical analog of Newton's classical equations of motion. The basic process of nature is thus forward in time, even though the "Instants" along which the collapses occur are not the non-relativistic "flat" surfaces, all points of which lie at the same time. The temporal advance from one global instant to the next can be confined to a small spatial region. This means, for example, that the first phase of Process 1, namely observer's choice of probing action, can be regarded as a local process, confined to a limited region, whereas the second phase, namely Nature's choice of response, turns out to be (see Appendix 1) non-local—i.e., faster-than-light!

The Role and Importance of Free Will

The linkage between the *free question* and the *random answer* ties the mental and physical aspects of things into a single cohesive dynamically evolving reality. In this evolution, the mental actions of observers play an essential role. Each Process 1 action (with positive outcome) leads to an increment in our collective knowledge, making your probing mind a causally efficacious active participant in the psycho-physical process that contributes to this collective/joint knowledge. Your "free choices" of probing actions, combined with nature's responses, enter actively into the determination of both your future psycho-physical states and the futures also of other observers of the system observed by you.

Rational arguments lead (see Chap. 10) from the explicitly dualistic form of Orthodox von Neumann Quantum Mechanics to the conclusion that all aspects of Nature, including our own mental aspects, must be interacting parts of one "mental whole". Understanding oneself to be an integral part of a "mental whole" tends to elicit a feeling of connectivity, community, and compassion with fellow sentient beings, whereas the materialist notion of mechanical action and survival of the fittest tends to foster disregard of the feelings and welfare of others.

One's entire approach to life tends to rest on whether one views oneself as an efficacious component of a "mindful whole", or a tiny cog in an essentially mindless machine, with a mysteriously attached but physically powerless mind that pointlessly spins false delusions about its physical power.

A classical mechanistic powerless self-image can have a tendency to produce attitudes of resignation, depression, hopelessness, pointlessness, and amorality. However, the quantum self-image, which makes your mental valued-based efforts causally effective, tends to create a more dynamic, elevated, hopeful, forward-directed, moral attitude. Recent experiments by psychologist Jonathan W. Schooler and others reveal a positive empirical correlation between people's belief in free will and the morality of their actions. Quite generally, your attitude and actions depend strongly on your beliefs about yourself in relation to the reality in which you are embedded. In today's educated world, your beliefs about these matters are likely to depend strongly upon what you believe science says you are.

Chapter 5
The Physical Effectiveness of Conscious Intent

A Brief Review

Classical mechanics says that your thoughts cannot affect the behavior of particles; but realistically construed orthodox quantum mechanics says they can.

The original Copenhagen quantum theory was designed by its founders merely to reliably predict relationships between what we do—the experiments that we perform—and the outcomes that we then observe. Thus, even at this initial stage of quantum mechanics, we observers do not just passively witness; we also purposively act. We choose, on the basis of our personal values, what our atom-built brains and bodies will do. These actions are often tied to our imagined perceptions of perceived events that might occur "out there", and in the future.

But when formalized by the work of von Neumann, the theory produces a dynamical conception of reality that can automatically–explain—without any change in the dynamical rules—not only the statistical relationships between our probing actions and our resulting perceptions, but also the capacity of a person's mental intentions to influence that person's bodily actions in the way that he or she mentally intends. That is a significant development: an automatic explanation of how a person's free-willed intentional mental effort can influence, in the intended way, that person's bodily actions. Quantum mechanical dynamics thus joins together what classical mechanics has rent asunder: our minds and our bodies.

According to the standard quantum principles, our choices of which probing actions to perform, and when to perform them, are not determined by any quantum mechanical law. These choices enter the theory as *free (un-coerced-by-material-causes) choices*. And a suitable structuring of those free choices can tend to make your physical body move in accord with your mental intent.

How can this come about? How can something as intangible as a mental intention, cause physical objects such as your arms and legs to move in intended ways? By what process can the motions of your fingers come to express the complex thoughts that you intend to express in written words?

© Springer International Publishing AG 2017
H.P. Stapp, *Quantum Theory and Free Will*,
DOI 10.1007/978-3-319-58301-3_5

The Quantum Zeno Effect

Within the Orthodox Quantum Mechanical description of nature, this physical power of your conscious thoughts can arise from a well-known rigorous property of quantum mechanics known as the Quantum Zeno Effect, and sometimes as the Anti-Quantum Zeno Effect.

Suppose a physical/material system is being probed by an observer whose mental aspect, his ego, is free to choose a sequence of probing Yes-No questions that will elicit responses, Yes or No from nature. And suppose this ego would like the observed system originally observed to be in state "Phi(0)" to move from that state to a perceived final state $\rho(1)$ along a smooth path $\rho(t)$ as t changes from t = 0 to t = 1. Then the quantum mechanical laws of motion entail that if the ego chooses to pose at each time n/N in the set of times {1/N, 2/N, 3/N, ... n/M, ..., N/N} the question "Do I perceive the observed system to be in the state $\rho(n/N)$?", then the probability that all N responses of nature will be "Yes" tends to unity (i.e., one) as N tends to infinity: all of nature's responses will, by virtue of the Born Rule, almost surely be "Yes" if N is sufficiently large. Thus by a suitably rapid choice of probing questions the observer can, by its choices of these questions, effectively control both the perceived responses and the associated material reality that is being perceived.

Because the observer's choices are stemming from the mental realm, there is no known limit on exactly how rapid these free choices can be. But it is reasonable to suppose that survival consideration make this effect far easier to use if the action is directly an action on the observer/actor's sensitive brain than on a perceived brute external system.

Thus the behavior of the brain, according to Orthodox Quantum Mechanics, is not completely determined by prior physically described properties of the universe alone, but can be significantly influenced by "free choices" made by human observers pertaining to which probing action to instigate, and when to do so. Here, again, the "free" in "free choice" means, specifically, that this choice is not determined by prior physically-described aspects of the universe alone. Our conscious free choices and mental efforts enter naturally, according to the quantum mechanical dynamical laws, into the evolution of the psycho-physical universe.

I shall describe next how, within Orthodox Quantum Mechanics, the simple holding-in-place action produced by the Quantum Zeno Effect can tend to make a person's physical actions conform to that person's mental intent.

Templates for Action

Suppose, for example, that you are struggling unsuccessfully to put a heavy object onto a truck in order to make needed home improvements. Suppose you are mentally wrestling with whether to try harder, get help, or give up.

It is reasonable to suppose that in this situation your brain will construct, via bio-physical processes, and on the basis of previously learned habits and responses, several different patterns of neurological activity, each of which would, if held in place for a sufficiently long period, while the other patterns are suppressed, send out a sequence of neural impulses that would cause your body to behave in one of the possible ways responsive to your plight. Such a neural pattern is called a *"template for action"*.

This situation can be analyzed within either orthodox QM or classical mechanics. If one uses classical mechanics then the fact that in classical mechanics our minds enter only as passive observers means that that theory can give no dynamical explanation of the connections between, on the one hand, your *feeling* of making a mental choice/effort to actualize some intended bodily actions, and, on the other hand, the associated responding movement of your physical body—for your mind is required to be causally inert: not part of the dynamics. But why should one's mind exist if it is not logically entailed by the classical material activity, and has no function? And why should it delude us into believing it is causally effective if it is really doing nothing? And how can we actually construct a purely (mind-independent) material dynamics that gives the same predictions as the empirically successful combination of Processes 1 and 2 that brings our known-to-exist minds into the dynamics in a way that conforms to how it feels to behave as we do, and that is in harmony with the evidence of every-day experience?

Because we know that our thoughts and mental efforts exist, and hence probably have an important function, is it not an irrational tour de force to try to show that they exist yet have no causal power? Why should we try to evade using such a wonderful theory that is so well defined, both mathematically and logically, and that works so well in practice, instead of trying to more fully exploit it?

If one simply adds, ad hoc, to classical mechanics the *postulate* that certain brain activities "produce", or "are", the associated conscious thoughts, then a correlation between brain activities and consciousness is *imposed by a fiat having no logical roots in the classical physical theory*. It causes our consciously instigated probing actions to become postulated *effects* of physical brain actions, not their *causes*, as in quantum mechanics, and it consequently reduces us to effectively thoughtless 'mechanical automata'.

But why is this seventeenth century notion, which is so despised by so many philosophers, and which is so contrary to our first-hand feeling of what we are—and is so seemingly absurd and senseless—be such a compelling desiderata today, when it has been reversed by the superseding hugely successful contemporary quantum physics? This current theory recognizes our experience-based knowledge as both the proper foundation of science and also an essential part of the cause of the Process-1 choice of probing action. Why should this reasonable, useful, and well-defined role specified by contemporary physics be rejected in favor of what seems a total absurdity about our minds proclaimed by a failed seventeenth century false start that makes our lives a farce?

The quantum alternative is a mind-matter dualism in which our minds, by virtue of their capacity to freely choose probing questions, **combined with nature's Born-Rule-restricted reply,** allow us to tend to actualize the bodily actions that we mentally intend! Our minds do matter in a way essential to our physical survival, well being, and purpose.

Chapter 6
Reality and Spooky Action at a Distance

Spooky Action at a Distance

Quantum mechanics has a peculiar feature that Einstein called "Spooky action at a distance", and which he found problematic. The problem arises under certain realizable empirical conditions involving two different experiments performed at essentially the same time in two far-apart experimental regions. Under the specified empirical conditions, the computational rules of quantum mechanics assert that the psycho-physical event initiated by the "free choice" of experiment made in one experimental region instantly changes the quantum state that controls the quantum predictions about outcomes appearing in the faraway region. The empirical validity of these predictions has been amply confirmed by experiments of a kind first performed in the early 1980s by French experimentalist Alain Aspect and his colleagues, and that are by now quite commonplace.

The key phrase "free choice" means that the choice of which large-scale measurement procedure is performed in the region can be selected whimsically by the experimenter, or by a quantum random number generator, or by any other process that is effectively uncorrelated to the system being measured. The essential point is that the quantum prediction for what will be consciously perceived depends directly upon which experiment is set up and performed, but not upon the manner in which that experimental setup is chosen: the process of choosing the experiment is required/presumed to be dynamically independent of the system being probed by the experiments. This lack of dependence is generally agreed to mean that the choice that emerges from the process of choosing the experiment can be treated, in the analysis of these experiments, as a locally generated free variable.

© Springer International Publishing AG 2017
H.P. Stapp, *Quantum Theory and Free Will*,
DOI 10.1007/978-3-319-58301-3_6

The Quantum Dynamical Origin of "Spooky Action"

The "spooky action" arises within the quantum mechanical formalism from the fact that the quantum state of the entire physically described universe is defined at each instant of the advancing sequence of instants of time (or at a relativistic generalization of an instant of time) at which a "collapse" occurs. And this state represents an objective (collective) state of the known physical state of affairs over all of 3D space at that instant of time. If, at some instant, nature makes a choice of response to a probing action that is localized in some confined spatial region then, according to the basic quantum rules, the quantum state of the universe changes not just in that local region, but over all of 3D space at that instant of time. This abrupt global change is called a "collapse" or a "reduction" of the quantum state. Einstein called this *global collapse* "spooky action at a distance", and believed that no such nonlocal action could be physically real.

Evasion via Pragmatism

Theoretically explicit and essentially instantaneous actions at a distance occur in quantum mechanics in conjunction with the collapses/reductions of the quantum state. These collapses are essential features of practical quantum mechanics: They are needed to keep the quantum state in line with our empirical knowledge. But the existence of such transfers of information conflicts with Einstein's theory of relativity, which limits the speed of the motion of physical matter, and, correspondingly, the speed of any transfer of information, to the speed of light.

The founders of quantum mechanics did not want to admit or suggest that, in defiance of the theory of relativity, information could *really* be transmitted faster-than-light. Hence they evaded the problem by adhering to the pragmatic position that the "physical state of affairs" represented by the quantum state was not a representation of physical reality itself, but something more akin to human knowledge than to classically conceived matter. Being also unwilling to defend the idea that the physical state represents some absolute kind of knowledge, which would mean a retreat to an "idealism" deemed antithetical to science, the founders adopted the evasive position that the quantum mechanical state was merely part of an invented practical human tool for making predictions about upcoming empirical findings. Thus no claim was made that the quantum mechanical state represented "reality"; no claim was made about any "real" property of nature itself! Direct conflict with Einstein's ban on "real" faster-than-light transfer of information was thereby dodged.

The EPR Paper

The *assumed* absence of any real "spooky action at a distance" was the basis of an effort by Einstein to prove that a quantum mechanical description could not provide a complete description of physical realty. In 1935 he wrote, in collaboration with two young colleagues, Boris Podolsky, and Nathan Rosen, one of the most renowned scientific papers of all time [6], entitled: "Can Quantum Mechanical Description of Physical Reality Be Considered Complete?" This paper is usually identified by the initials of the last names of its three authors—'EPR'. It assumes, in concordance with the theory of relativity, that information cannot be transferred faster than light, and then argues, on the basis of the predictions of quantum mechanics, that this theory cannot provide a complete description of physical reality. But the authors opine that a complete (and non-spooky) theory of physical reality is possible.

Bohr's Response to EPR

The easy response to EPR by the founders would have been to simply re-emphasize that quantum mechanics does not claim to describe physical reality itself. However, a simple response of that kind would have sparked, among scientists who aspire to be more than high-level engineers, efforts to find a more complete theory. Making such efforts is exactly what Einstein believed scientists interested in basic questions ought to be doing, but what the founders of quantum mechanics believed that potentially useful scientists *ought not* to be doing. Thus Niels Bohr, the senior founder of quantum mechanics, chose to answer EPR by focusing on the slippery question of what constitutes physical reality.

What, exactly, is "physical reality"? A logically sound argument pertaining to "physical reality" requires giving some definite meaning to that phrase. But our ideas about physical reality are deeply influenced by our experiences of the world around us, which *seem* to conform to the principles of classical physics. Thus, any proposed characterization of physical reality is in jeopardy of being challenged as resting on intuitive classical ideas alien to the quantum precepts, and hence as being prejudicial: as begging the question.

The EPR paper was built, therefore, *not* upon some notion of "physical reality" that could be attacked as obscure, unscientific, or question-begging. It rested, instead, on the *demand*—enshrined in Einstein's theory of relativity—that information cannot be transmitted faster than light. The opening for using this demand was slipped into their famous "Criterion of Physical Reality". This criterion asserts that "If, without in any way disturbing a system, we can predict with certainty (i.e., with a probability of unity) the value of a physical quantity, then there exists an element of physical reality corresponding to that quantity." The requirement "without in any way disturbing" was met by considering situations in which the possible disturbance would require faster-than-light action. EPR were then apparently able to conclude that a certain pair of properties (that were represented by non-commuting operators) were *both* physically real, and hence simultaneously definable, although the principles of quantum mechanics are unable to encompass that possibility. Thus the quantum mechanical description was proved to be incomplete.

Of course, a simple alternative conclusion would be that faster-than-light actions *can* occur!

Most of the EPR argument was straightforward physics and not open to challenge. But it depended upon one metaphysical element, the EPR Criterion of Physical Reality, which begins with the words, "If without in any way disturbing a system ..."

Bohr [7] attacked this metaphysical element of the EPR argument in a subtle way. Bohr states:

> Of course there is in a case like that just considered no question of a mechanical disturbance of the system under investigation during the final last critical stage of the measuring procedure. But even at this stage there is essentially a question of *an influence on the very conditions which define the possible types of measurements regarding the future behavior of the system.* Since these conditions constitute an inherent element of the description of any phenomena to which the term 'physical reality' can be properly attached, we see that the argumentation of the mentioned authors does not justify their conclusion that the quantum mechanical description is essentially incomplete.

Bohr argued, however, that quantum mechanics was *pragmatically* complete, which, in the end, is what matters most to most physicists, who could now, if challenged about the failure of science to talk about physical reality, refer to Bohr's reply to the EPR argument pertaining to that issue.

Notice that the EPR argument is based on the matter-related assumption that, in physical reality, information cannot be transferred faster than the speed of light; whereas Bohr's argument is based on the *pragmatic* idea that our understanding of nature should be based, not on prejudicial presumptions about imagined–to–exist matter, but on our actual knowledge, and on the possibilities of our future knowledge. So the conflict comes down to the question of the proper foundation of science: Is it the materialistic concepts of the classical physics stemming from the postulates of Isaac Newton? Or is it what we actually know, or are able to know, as was urged by David Hume and the other empiricists.

In spite of this fundamental disagreement, the two protagonists did agree on one key point: There could be no *real* transfer of information over space-like intervals. But that presumption is proved wrong under the weak conditions of the proof given in Appendix 1. That proof rules out Einstein's classical-matter-based conception of physical reality, but is completely compatible with the psycho-physical conception of reality specified by von Neumann's orthodox formulation. There is no need to retreat from the idea that a rationally coherent basic realistic physical theory, namely realistically construed orthodox QM, can accommodate: (1), the findings of atomic physics; (2), the classical character of appearances; (3), the evidence for the causal effectiveness of our mental intentions; and (4), "spooky actions at a distance".

Bell's Theorem and the Nature of Reality

Historically, this controversy lay semi-dormant, with practicing physicists generally siding with the pragmatic position of Bohr, until John Bell wrote a paper [8] based on the notion of "hidden variables". Bell's hidden-variable approach added to the usual assumptions of the validity of the empirical predictions of quantum mechanics, and the notion of effectively "free choices" on the part of the experi-menters, the further assumption that there is an underlying invisible "hidden" physically described substructure that carries the causal connections. In exact analogy to classical statistical mechanics, each empirical situation is represented by a sum of statistically weighted physically defined possible "real states of the system being studied. Following Bell's lead, these quantum analogs of the physically defined possible "real" states of classical physics are labeled by the Greek letter "lambda".

This hidden-variable quantum theory is applied in the context of the (Bohm-Bell) experimental situation, which involves two spin-1/2 particles existing initially in a certain (e.g., singlet) spin state. These two particles then fly apart in opposite directions to two far-apart experimental regions, in each of which a measurement is performed. These two far-apart measurements are performed at essentially the same time in the common rest frames of the two experiments. The measuring device in each region has two alternative possible settings. So, alto-gether, there are four settings under consideration and for each setting, two alter-ative possible outcomes.

One of these alternative possible outcomes is labeled with an identifying label "plus one" and the alternative possible outcome of the same measuring process is labeled with a "minus one". (I am assuming here, for simplicity, perfect geometry and 100% efficient particle detectors, and shall stick to this idealized case).

Each possible real state lambda of the universe specifies which one of the two alternative possible measurements is performed in each region, and, for each of these two possible measurements, which one of the two alternative possible out-comes of that measurement occurs. The choices of which measurements are per-formed in the two regions are treated as two free variables, and the "no-faster-than-light-transfer-of-information condition" (No-FTL) is imposed by requiring the outcome in each region to be independent of which measurement is chosen and performed in the far-away region.

Bell's analysis is based on a "correlation function". This function specifies, for each of the four considered pairs of settings that include one setting in each region, the "degree of correlation" between the labeled co-occurring outcomes in the two regions. This correlation function is a number that can vary from the value plus one (if the labels of the co-occurring outcomes in the two different regions always agree) to the value minus one (if the label of the occurring outcome in one region is always opposite to the label of the co-occurring outcome in the other region. This function is defined, for each pair of settings of the devices that has one setting in each of the two regions, by averaging the product of the labels of the co-occurring outcomes in the two different regions. Here "averaging the product of the labels" means summing over the product of the four alternative possible combination of paired labels, $\{(1, 1), (1, -1), (-1, 1), (-1, -1)\}$, dividing by 4, and weighting each pair with the quantum probability of that possible combination of the two far-apart outcomes, for the specified-by-lambda pair settings of the two devices.

The key step of Bell and associates is to invoke the demand for no faster-than-light-transmission-of-information (No-FTL) by converting, for each fixed lambda, the contributions to the correlation function into a product of two separate factors, each containing the dependence on both the choice of setting and the choice of outcome in just one of these two factors.

Thus No-FTL is implemented by Bell by factorizing the formula for the correlation function! But this factorized form cannot be simultaneously valid for a certain four alternative possible choices of the pair of settings in the two regions. Thus implementing NO-FTL in this way leads to a contradiction.

If this factorized form were to be acceptable, then the hidden-variable theory could, with some justification, be said to represent a certain "local realism". That title, "local realism", is the title that its proponents seem to prefer, to "local hidden-variable theory", in conjunction with their claim that "No local realistic theory can be compatible with the predictions of quantum mechanics." But that wording invites drawing certain logically unwarranted conclusions.

The hidden-variable theory can reasonably called "realistic": it is basically similar to classical statistical theory, which rests on the normal classical idea of real material worlds. If a certain "locality" condition is then imposed, and contradictions with empirical data and quantum predictions ensue, then it might seem that, given a demand that the basic theory must describe reality, nature must be nonlocal, contrary to Einstein's demands.

However, there are other ways to achieve a realistic ontology, and one of them might be able to evade the need for non-locality. Hence FTL is not entailed merely by the need for it in the hidden-variable model.

Moreover, it has been well-known since a 1984 paper of J. Jarrett that factorization is equivalent to the conjunction of "parameter independence" and "outcome independence" (in the terminology of A. Shimony). Parameter independence is the same as No-FTL transfer. So the failure of the factorized formula to accommodate the quantum predictions (and the empirical data) entails either FTL transfer or "outcome dependence" (the dependence of the outcome in one region on the "outcome" in the other region). But "outcome dependence" is the normal feature of the empirical predictions in the situations under consideration here. Hence one cannot conclude from the failure of factorization that there must be FTL transfer of information: the failure of outcome independence suffices to account for the failure of factorization.

An actual proof that there must, under appropriate conditions, be FTL transfer of information is given in Appendix 1. It achieves what the hidden-variable approach may seem to claim to achieve, namely the need for FTL transfer of information about experimenter free choices, but does not logically achieve.

Chapter 7
Backward-in-Time Causation?

Orthodox Quantum Mechanics is based on the idea that a physically described state of the universe exists at an instant of time over all of three-dimensional space, and advances, event by event, into an indeterminate future, leaving behind a fixed and settled sequence of past states. Certain phenomena associated with this chain of events appear to involve backward-in-time causation, but they are accommodated without introducing any actual backward-in-time action. It is important for a valid understanding of nature to understand how orthodox quantum mechanics accommodates seeming backward-in-time actions.

Delayed Choice Experiments

John Archibald Wheeler [9] described an experiment that seemed to show that an experimenter's "free choice" about which experiment he or she performs at one time can affect what happened at an earlier time. The essential point can be illustrated by the following idealized version.

Suppose you have super-sensitive vision and can detect individual photons falling upon your retina. Imagine that you are looking through one eye at a screen with two small holes, through which light of a visible frequency is moving in your direction. Quantum Mechanics says that if you focus your vision on the screen, and the light is sufficiently weak, and your vision is sufficiently sensitive, you will see the individual photons passing essentially one at a time through either one hole or the other. But if you choose to focus, instead, on a location far behind the screen then the photons will still come one at a time, but will build up a complex interference pattern that depends on the distance between the two holes, apparently showing that the light associated with each individual photon has, in some sense, passed through both holes. Thus, what happens earlier at the screen, namely the individual photons passing through both holes, or passing through just one hole, seems to depend on your later choice of how to focus your eye.

© Springer International Publishing AG 2017
H.P. Stapp, *Quantum Theory and Free Will*,
DOI 10.1007/978-3-319-58301-3_7

Essentially the same experiment can be performed with devices that act so fast that the choice between the two alternative possible focal lengths can be made by a random number generator <u>after</u> the photon has passed through the screen. Thus it would appear that, in some sense, the photon either passes exclusively through one slit or the other, but not both; or, alternatively, through both together, depending on which kind of observation is chosen *after the photon has already passed through the hole or holes.*

This kind of experiment is called a "Delayed Choice" experiment, and various refinements of it have been designed and successfully carried out by Scully and colleagues [10]. The observed phenomena certainly conform to the just—described predictions of quantum theory, but the 'causal implications' need further discussion.

The "Bohmian" Approach to Explaining 'Causal Implications'

For example, one proposed way to understand quantum mechanics was advanced in the early days of quantum mechanics by physicist Louis de Broglie. It was then pretty much abandoned due to criticisms by Pauli, but resurrected and developed by David Bohm [11] in the 1950s. This way of understanding the success of quantum mechanics asserts that there really is a classical-type world of tiny particles, but *also* a wavelike quantum state of the universe that evolves *always* in accordance with the Schrödinger equation, and hence never "collapses" in association with an increase in "our knowledge", as specified by both the Copenhagen and Orthodox versions of Quantum Mechanics. In Bohm's no-collapse Quantum Mechanics the function of the wave is to "guide" the particles, which are assumed to be the aspects of Nature that control our conscious experiences.

In this "Bohmian" model of reality the changes made in the focusing of your eyes influence the evolution of the quantum wave within your eyeballs, and this change in the wave, which travels through both holes, influences the trajectory of the photon (particle of light), which travels though only one hole, *when it lies inside your eye, which is focused in one way or the other.* This theory correctly accounts for the delayed-choice phenomena without invoking, in our two-slit experiment, any notion of backward-in-time action or 'causation'. The difference in what is observed is due to the classically understandable causal effect upon the trajectory of the photon on the way the eye is focused—or in other versions of the experiment, on the details of experiment determined at the last moment by an independent random number generator (RNG).

This Bohmian approach does account for the physical properties, considered as self-determining physical properties. But the classical demand that these particles interact essentially by contact interactions in ordinary 3-D space is grossly violated in the two-observer FTL experiment. In that experiment, which experiment an experimenter in one spatial region decides to perform can have a big effect on which outcome appears at essentially the same instant in a very faraway laboratory. And, as in classical physics, the theory says nothing about our minds.

If "science" is properly about our growing knowledge, then the theory is fundamentally incomplete, for it offers no rational account of how the physical description that it provides is tied to our evolving mentally described knowledge, and how, if at all, our mental intentions influence the activities of our brains, and hence bodies.

Bohm himself addressed this problem, and was forced to replace his original simple non-local theory by a very complex one in which the mystery of mind is transferred to a mystery about an infinite tower of observing systems, each observing what was physically happening in the level just below. The simplicity and attractiveness of his original quasi-classical quantum theory was lost when he tried to incorporate our human experiences. But incorporating our causally effective mental intentions is exactly what realistically construed orthodox Quantum Mechanics achieves! It offers a rationally coherent mathematically formulated description of reality that includes an account of how our mental intentions influence our physical behavior in a way that is concordant with all the empirical evidence, instead of defaming us by claiming our most important human quality, the capacity of our mental intentions to influence our bodily actions, an illusion or delusion. The price to pay for this increase in rational understanding is a failure of Einstein's classical-physics-based intuition that information cannot travel faster than light. But in quantum mechanics one has, instead, the property that no "signal" (sender chosen message) can be sent faster than light. This weaker property of quantum mechanics suffices to maintain in the quantum world the essential empirical requirements of the theory of relativity.

Chapter 8
Actual Past and Historical Past

The evolving history of the universe is normally regarded as being divided into three parts: past, present, and future. The present instant "now" separates the past that has already happened from the future that has not yet happened. One idea of the nature of things is described by the phrase "closed past, open future". It indicates that the past is already fixed and settled, whereas the future is not yet determined. Another idea is the "block universe" in which every event in the entire history of the universe is already pre-determined: is already fated to be exactly what it was, or will eventually turn out to be.

Deterministic classical mechanics is usually regarded as defining a "block universe". Einstein considered the universe to be a block universe, and his theory of special relativity dealt with many different ways that one can assign 4 coordinates (x, y, z, t) to the space-time points in this block universe in which the entire pre-determined course of the history of the universe is laid out: there is no real "becoming".

Orthodox Quantum Mechanics, on the other hand, is based on a forward-in-time process consisting of a well-ordered sequence of psycho-physical events that are associated with a well-ordered sequence of "instants" "now", each of which is a smooth 3D surface in the 4D space-time. Each such surface is later at some point, but earlier at no point, from its immediate predecessor, as already discussed in Chap. 3.

Associated with each present instant "now", labeled by "σ", there is a state (density matrix) of the universe "$\rho(\sigma)$" that defines, via the "Born Rule", the probabilities of the various alternative possible outcomes of the probing action/question posed by the observer associated with the collapse event located on that present instant "σ". The state of the universe, $\rho(\sigma)$ can be called the "actual past" of the instant σ: it is the immediate past that partially controls, in conjunction with the observer's probing question, the collapse occurring on the instant "σ".

This collapse event has, according to the orthodox *collapse* interpretation, changed the immediate future potentialities by eliminating from them all causal effects that stem from those parts of the prior quantum state that were eliminated

© Springer International Publishing AG 2017
H.P. Stapp, *Quantum Theory and Free Will*,
DOI 10.1007/978-3-319-58301-3_8

by the collapse that occurred at the instant "σ". The new set of immediate future potentialities is, because of this collapse that occurred at the instant "σ", different from what it had previously been. Consequently, the *"actual past"*, does not provide the pertinent-to-the-future *representation* of the past.

The representation of the past that is pertinent to the future is the one that smoothly evolves, according to the continuous laws of motion—the Schrödinger-like equation—into the quantum state that has just been created. This *causally pertinent state of the past* is called the *"effective past"*. It is specified by evolving the newly created state backward in time, by means of the inverse of the pertinent Schrödinger-like equation. Thus, the *"actual past"* is the state of the universe that existed just prior to the present instant "now", whereas the *"effective past"* is the part of that past state that smoothly evolves into the immediate future, and therefore pertains to the potentialities associated with the next instant. Between the two instants the state evolves via Process 2.

The *"effective past"* contains, in particular, the <u>records</u> of those parts of the past that have survived the recent collapse, and are thus *relevant to the current future*. Since the "effective past" contains these surviving records, it is the part of the actual past that we can in principle recall. The rest of the actual past is eradicated by the collapse, and is forever non-recoverable.

These important aspects of quantum mechanics are succinctly captured by an assertion made in the recent book "The Grand Design" by Stephen Hawking and Leonard Mlodinow: "We create history by our observations, history does not create us" [12].

We, by our free (asserted to be un-coerced by matter) choices of our probing questions, influence the evolving form of the material universe by means of the FTL effects of these choices on a sequence of global (nonlocal) collapse events.

The rules of relativistic quantum field theory (RQFT) ensure, however, that these instantaneous actions can never be used to send a *sender-controlled message* "faster-than-the-speed-of-light" *to* an *actual receiving intelligence*. Hence Einstein's demand of "no-faster-than-light transfer" of information is in fact satisfied in orthodox QM insofar as this "information" is "sender-controlled information" that becomes *known* to a receiving observer.

Thus, insofar as one accepts that we live in a real quantum (not effectively classical) universe described by orthodox quantum mechanics, in which our choices of our probing actions are not coerced by material processes, care must be taken to make distinctions that have no counterpart in Classical Mechanics – for example, whether the effect is on the consciousness of an actual person or upon quantum potentialities for a collection of four alternative possibilities at most one of which can actually happen.

To see how these considerations play out let us consider the "Wheeler delayed choice experiment". At the moment that the pulse of light is passing through the holes, the quantum wave that represents the light is divided between the two holes.

If, at a later time, the observer sees the photon coming through the left-hand hole then, according to the rules of orthodox quantum mechanics, a global collapse will occur: the parts of the quantum state incompatible with that experience will be obliterated. The new state, representing the potentialities for the future experiences of all observers, will be the continuation into the future of the surviving part of the prior state. *The continuation of the new state backward in time, using the Schrödinger equation in reverse, is the effective past.* The existing evolving state is, as Hawking and Mlodinow state, created by our observations (together with nature's responding actions). All future experiences of all observers will be concordant with the empirical fact that the photon was seen by some observer to pass through, say, the left-hand hole.

The evolving situation during the time that the pulse of light was passing through the screen was that the wave was passing through both holes. That fact is fixed and settled: it is never revoked. But if the observer poses the question "Do I see the light coming through the left-hand hole", and Nature's response is "Yes", then the quantum state collapses to the part compatible with the observer's experience. This state, extended backward in time via the Schrödinger equation acting in reverse, will have the light wave passing through the left-hand hole, and the effect of this observation will be incorporated into all future experiences of all observers. All of this is logically captured and mathematically represented by von Neumann's conception of the nature of things.

This orthodox way of understanding the apparent backward-in-time effects uses only strictly forward-in-time evolution of the quantum state. It achieves an explanation of an apparent retro-causation by using the orthodox forward-in-time dynamics.

Some of these rules lead to the continual generation of "smears" of alternative classically conceivable, but mutually incompatible, possible worlds. Other orthodox rules govern the collapses of this evolving quantum state. These collapses systematically trim away the branches of this growing quantum state that become irrelevant to the future. These branches have become irrelevant because they led to possibilities that were probed by observing actors/agents, and were eliminated by Nature's choice of reply.

Each such Nature-produced collapse, although precipitated by the probing action of some localized observer, is a global event. It instantly alters aspects of the quantum state that pertain to observations about to occur in faraway regions. This instantaneous effect ("spooky action at a distance") is incompatible with the no-faster-than-light-transfer conditions asserted by Einstein's Theory of Relativity.

By proving that these "spooky actions at a distance" are unavoidable consequences of a few well-verified predictions of quantum mechanics *pertaining exclusively to macroscopic phenomena*, without introducing any conditions pertaining to microstructure, the theory rules out the possibility that the world of macroscopic phenomena can be rationally understood as being built out of classically conceived matter. Materialism is ruled out.

The important implication of this analysis is that it rules out a notion prevalent among philosophers and scientists interested in the mind-matter connection that quantum effects pertain only to microscopic processes, and that, apart from the elements of quantum randomness, the quantum character of reality somehow magically disappears at the level of macroscopic physical processes. There is no science-based justification to believe that the behavior of the brain of a conscious person can be understood in terms of the concepts of classical physics: The "promise" in what Sir Karl Popper called "Promissory Materialism" conflicts with contemporary basic science.

Chapter 9
The Libet "Free Will" Experiments

According to nineteenth century classical physics, reality is described in purely physical terms, and is deterministic: the world is described in terms of mathematical properties attached to space-time points, and *the future is completely determined by the past*, by virtue of mathematical conditions on these properties. This formerly-believed feature of nature is called the "Causal Closure of the Physical").

In stark contrast, the *purely physical aspects* of the quantum laws determine from the physically described aspects of the past only a "statistical mixture" of future potential physical worlds. Thus the purely mechanical aspects of the quantum mechanical laws of motion have a "causal gap": they *do not determine* from the physically described aspects of the *past* the physically described aspects of the *future, but only a quantum statistical mixture of such aspects*. 'Something else', or 'something more', must enter into the causal structure in order to determine the 'nature of reality' under consideration at this specified point in time. This *lack of determinateness* (inherent in quantum mechanics at this stage of determining the 'nature of reality') is not resolved simply by specifying the value of some "quantum element of chance".

According to the basic precepts of standard ("Copenhagen-von Neumann") Quantum Mechanics, *a reduction of the uncertainty represented by this quantum statistical mixture requires* that a particular *probing action*, specified by a 'Yes/No' question, be chosen by an observer. Furthermore, an answer, 'Yes' or 'No', is required to be chosen and reified (made concrete or 'actualized') by Nature. As a result, two key questions arise:

(1) What determines which 'Yes/No' question the observer will choose? and
(2) What determines Nature's answer, 'Yes' or 'No'?

The answer to the second question is that the answer, in the form of the binary 'Yes' or 'No', is determined by the 'infamous quantum element of chance'. But what determines which 'Yes/No' question the observer will choose?

© Springer International Publishing AG 2017
H.P. Stapp, *Quantum Theory and Free Will*,
DOI 10.1007/978-3-319-58301-3_9

In actual scientific practice, the answer to this question is *determined by an observer's personal choice* of what to attend to. *This choice is definitely <u>not</u> determined by the quantum physical laws, and it is, in that specific sense, <u>a "free choice"</u>*. This choice is thus 'naturally determined' (within the "Copenhagen-vN" framework) by the *observer's mental aspect*, and hence from values embedded in the *observer's mental aspect, his "ego"*.

These quantum mechanical precepts can be illustrated by showing how they work in practice, in the famous "free will" experiments performed by Benjamin Libet and his associates (Libet 1985)[1].

Introduction to Libet

We all feel that certain of our conscious thoughts can, and often do, *cause* our voluntary bodily actions to occur. Our lives, our institutions, and our moral philosophies are largely based on that intuitive sense of how the world works. In fact, the entire notion of "cause" probably originates in that deep-seated feeling or intuition.

An evidence-based argument *against this basic intuition*—that our thoughts can influence our bodily actions—stems from an experiment performed by Benjamin Libet and his associates (1985, 2003). In this experiment, a human subject is instructed to perform (voluntarily) during a certain time interval, a simple physical action, such as raising a finger. Libet found that a measurable brain precursor (the "readiness potential") of the "conscious choice" to promptly perform an action, occurs in the brain about one-third of a second prior to the occurrence of the psychologically experienced act of "willing" that action to promptly occur.

This empirical result appears, on the face of it, to show that the *conscious act of "willing"* must be a *consequence* of the associated brain activity, not the *cause* of it. For, according to the normal idea of *cause*, a "free choice" cannot *cause* a prior happening to occur.

This example is just one instance of a general feature of mind-brain phenomena, namely the fact that a conscious experience of choosing often seems to occur after a lot of preparatory work has already been done by the brain. This fact, combined with the classical mechanics precept of the "Causal Closure of the Physical", leads, plausibly, to the conclusion that the *felt causal potency of our conscious choices* is an 'illusion'.

---

[1] A reference to Libet's work can be found in the Supplementary Material.

Libet in a Quantum Mechanical Framework

However, an examination of this Libet experiment, viewed from the perspective of the quantum framework (developed in the mid 1920s by the founders of quantum mechanics to deal with observed physical phenomena, and cast into logically and mathematically rigorous form already in 1932 by John von Neumann), shows these Libet results to be in good accord with both (1), the quantum-postulated *freedom of those human choices* from physical coercion, and (2), the capacity of those *value-based intentional mental choices* to *influence the chooser's future* bodily actions in the intended manner.

Any dynamical theory, to be relevant to our experienced lives, must link the dynamics to our streams of conscious experiences. Quantum theory is built squarely upon the demand that this condition be met. It is a *psycho-physical dynamical theory* that has *causal connections between human brains and minds* built intrinsically into it, in a way that allows our conscious choices to be both *free of physical coercion* yet *causally potent* in the physically described world.

In spite of this obviously extremely pertinent twentieth century revision of the relevant physical principles, *contemporary neuroscience* and philos*ophy of mind* largely continue to base their quest to understand human consciousness on the inadequate nineteenth century classical mechanical conceptualization of reality, which contrary to standard quantum mechanics, leaves our consciousness completely out of the causal dynamics.

The Libet Causal Anomalies

We often resolve to act in some specified way at some future time, and then meet this commitment with great precision. The brain, recognizing in the sensed input a need for an appropriate action, immediately begins to build a template for such an action. According to the ideas of William James, this template can be activated by an act of "consent", on our part, at the later time when the future 'commitment' is to be fulfilled.

In the Libet experiment, an initial instruction is given to the subject to willfully perform an action at some future time, say within the next few minutes—an act of 'raising a finger'. In the Libet case, the instruction is imprecise as regards the exact time of the called-for action. Because of this latitude in the timing of the willful action, the physical brain will presumably construct a sequence of alternative "templates for action", each designed to produce the specified action at a 'different alternative possible time'.

For the earlier templates, the urgency is low and Nature's reply will more likely be "No" (it *doesn't* really *have to be done*, *yet*). In accordance with principles described in earlier chapters, all recorded traces of these early failed attempts will be banished from the realm of recorded possibilities. Finally a "Yes" response occurs, and that outcome is recorded. *But all potentialities that do not lead to the outcome "Yes" (that actually occurs), are eradicated by the earlier "No" collapse events. This leaves the record only of the flow of potentialities that do lead to the template for action that is actually manifested: no record of the earlier potentialities that failed to manifest will survive the earlier collapses. And the still-forming templates for now un-needed actions will probably also dissolve.*

The processes in play here *begin* with automatic actions in the brain, which (responding to inputs from the surrounding physical world) constructs appropriate templates for action. This process is accomplished essentially in the same way that a classical-physics-based neuroscience might suggest. Each such template, if held in place for a sufficiently long period, while all conflicting activities are annihilated, will send out a sequence of neural pulses that will (if Nature's consent is given) produce a physical action. That is what templates for action do, if held long enough in place.

The observer's ego, surveying what is going on in its brain, selects, on the basis of its values, a template for action (which has been constructed by the brain) and asks whether what it (the ego) is experiencing is the perception which signifies that the selected template has been actualized. A positive feedback will inform the ego that its choice of probing action has influenced its steam of conscious experiences in the intended way

Only the "Yes" responses are experienced or remembered. The brain records of the failed attempts are annihilated by the associated collapses: a mental experience associated with a "No" response can in general not even be defined.

Libet Conclusion

The quantum mechanical understanding of the mind-brain dynamical system explained and defended in Schwartz (2005), and further elaborated in Stapp (2005, 2006), accommodates and explains the ability of our conscious intentions to influence our physical behavior. This theory covers in a natural way the Libet "free will" data. It reconciles Libet's empirical findings with the capacity of our conscious intentions to influence our physical actions, without these intentions being themselves determined by the physically described aspects of the theory.

This *empowerment of the mental participation of the observer/individual perceiver* is achieved by exploiting a "causal gap" in the mathematically expressed laws of quantum mechanics. This "causal gap" is filled, in actual scientific practice, by *invoking the conscious intentions of the human participants*. This practical and intuitively-felt role of conscious intentions is elevated, within the quantum ontology, to the status of an ontological reality, coherently and consistently integrated into quantum laws.

The Libet experiments are discussed further in the next chapter.

Chapter 10
Questions and Answers About Minds

The foregoing chapters have elucidated a science-based view of reality that I call "Realistically-Construed Orthodox Quantum Mechanics". It is profoundly different from the essentially Newtonian conception represented by classical mechanics. A comprehensive comparison of the two views in the preceding chapters has shown classical physics view to be inadequate in its treatment of reality, compared to that of Realistically Construed Orthodox Quantum Mechanics. Nevertheless, that classical physics view of reality is still employed by most contemporary philosophers, psychologists, and neuroscientists working on the problem of the connection between the mind and the brain. I comment on this situation by giving brief answers to five questions that I was asked at a recent meeting.

Question (1):   "Why bring conscious thoughts into the dynamical laws as independent inputs instead of allowing all mental properties to be determined by physical properties as in classical physics?"

Answer: Francis Crick, co-discoverer (with James Watson) of the double-helix structure of DNA, was a leading figure in the movement to recognize consciousness as a respectable/acceptable subject for scientific study. His influential 1994 book, "The Astonishing Hypothesis: The Scientific Search for the Soul" contains the famous passage:

"You, your joys and your sorrows, your memories and ambitions, your sense of personal identity and free will, are in fact no more than the behavior of a vast assembly of nerve cells and their associated molecules." He maintained, moreover, that the laws of classical physics would be generally adequate in the study of consciousness, and that *quantum physics* would be needed *only for tiny molecular-scale technical details*.

Crick's close associate, Christof Koch, has become a leading figure in neuroscience, but has backed away from Crick's ideas to some degree. At a recent small meeting in Berkeley, prior to Koch's talk (one in a long lecture series on "Unsolved Problems in Vision"), I questioned Crick's claim that conscious thoughts "are in

© Springer International Publishing AG 2017
H.P. Stapp, *Quantum Theory and Free Will*,
DOI 10.1007/978-3-319-58301-3_10

fact no more than" brain behavior. Koch promptly ridiculed that idea: "There are no yellow patches in the brain when experiences of yellow are occurring!" The philosopher, John Searle, was also present at that meeting. John and I hammered away at the lack of any theory (understanding) of the connection between these two (now admittedly different) things—conscious thoughts and 'brain behavior'.

I also focused attention on a passage in the paper (on Integrated Information Theory (IIT)) that Koch was expounding upon, which said: "IIT takes no position on the function of experience as such". It takes no position on whether "experience as such" does anything. But in his verbal response, it became clear that his position (like that of most neuroscientists) is that it is the brain/body that is doing every physically described thing: that our experiences are idle spectators that are created by the brain but have no reciprocal 'reactive effect' upon it. Our experiences have no reason, within the classical conception of reality, to exist at all: no job to do.

However, the classical physics conception is known to be profoundly inappropriate, and pursuing it has led to a growing number of very difficult"Unsolved Problems in Vision". Ontologically-construed RQFT is a radically different conception of the connection between our experiences and our brains that accounts rationally, and in mathematical detail, for all well established empirical data from planetary dynamics to atomic physics. In that conception, our conscious experiences play a critically needed dynamical role that makes "your joys and your sorrows" and your "free will" into the causally effective mental-type realities that they seem to us human beings to be.

How scientists view these matters is not just a matter of words, for those views control the research in fields of neuroscience and cogitative science.

Question (2): "Is it not true that the interaction of observed systems with the surrounding environment will account for the emergence of classical appearances, and thereby eliminate the need for the extra process that orthodox and Copenhagen Quantum Mechanics introduce via the quantum collapse?"

Answer: The answer is 'No'—it's not true! The interaction with the environment leaves the quantum state of the system being studied in a "quantum statistical *mixture*", which is generally a continuous "smear" of classically describable possibilities of the kind that we actually experience, each with zero probability to be actualized. To select the experience that actually occurs, some discrete selection process is needed.

Copenhagen and Orthodox QM deal with this need by means of an ordered sequence of two-part reductions, each consisting of a Process-1 "free choice" of a probing 'Yes/No' question made by a conscious observer, combined with a "random choice" of the answer, 'Yes' or 'No', made by Nature. The "free choice" is "free" in the sense of not being fixed by the physically described aspects of the theoretical structure (either alone or in conjunction with the famous quantum "random" element, which is connected to"Nature's" choice). Without this two-part selection process, or some substitute, the theory would fail. In particular, the effects of the environment on the system of interest are not sufficient to account for which particular experience actually occurs!

Question (3):   "How do you account for results of the experiments of Benjamin Libet and others which show that an associated brain action, called the 'readiness potential', precedes the mental act of consciously willing one's finger to move?" The fact that this brain activity *precedes* the willful act has been used by some authors to claim that the experience of willfully causing the physical act is a <u>conse-quence</u> of the brain's causing the finger to rise, *not a cause* of that physical action. That view is in line with the physicalist theories of the mind, which claim that a person's mind is simply a feature of the physically determined activities of that person's brain."

Answer: In order for a person to decide to perform a particular contemplated action, there must be a brain representation of that contemplated action. In Libet's experiments, the subject is instructed to perform a simple motor task (e.g. raise a finger) at some future time (implicitly understood to be some unspecified time in the next few minutes). According to the orthodox theory, this input instruction initiates a brain activity of constructing potential "templates for action" for various *alternative possible* actions that meet the specified temporal conditions. The early phase of each alternative possible "readiness potential" is a consequence of the brain activity of constructing such a "template for action", which involves also an account of the projected experiential consequences of initiating this action.

It was emphasized by William James that a contemplated action does not actually occur until an "act of consent" is given. In the Orthodox QM account, the process of consenting (or of allowing the potential action to become a part of the experienced communal reality) is *initiated* by a Process 1 probing action. As stressed before, this probing action is not fixed by the known physical laws. An initiating input coming from some other source is required.

Libet, mistakenly from this quantum point of view, associated the rise of the readiness potential with a decision to act. Then, to rescue "free will", Libet was led to his idea of a "free won't": a later decision by the observer that can override the supposed prior decision to act. But, according to the quantum model, the early part of this rise is merely a concomitant of the process of constructing a "template for action" that will *only later, by virtue of a mental choice,* be picked out from among the many potential templates that have been constructed in parallel by the Schrödinger-equation-controlled evolution of the quantum mechanical state of the brain of an observer. As explained in the chapter on apparent backward-in-time action, the only one of the parallel construction processes that will leave a record will be the one leading to the template selected by Nature to be actualized. The records of the others are destroyed by the collapses. This explains the rise of the readiness potential *before* the subject's free choice of probing action that produces—subject to Nature's positive response—the physical action that actually occurs. The "free choice" by the observer of what to observe and when to observe it thus enters in an essential way into the course of physical events. Our effortful mental intentions thereby become causally effective in the physical world.

Question (4): "Since mind is elevated to a basic role in your quantum view of reality, how do you distinguish those views from Western idealists such as Berkeley and from Eastern Philosophies based on Buddhist and Hindu teachings?"

Answer: All of these views arise from the empiricist premise that our understanding of reality should be *based on the structure of the realities that we really know exist*, namely our *streams of conscious experiences*. Since these various views all start from a mental foundation, and seek to produce a rationally coherent parsimonious narrative concordant with the physically described character of what we see around us, it is not surprising that they all arrive at somewhat similar conclusions. But the Eastern versions are more intertwined with Indian ideas of Karma, reincarnation, and lore than the Western versions that evolved in the context of Greek thought, Christianity, and the rise of science.

Question (5): "If mind is an important aspect of reality, then what do you say about the world before life emerged?"

Answer: I was asked this same question by Heisenberg, in his solicited comments on my 1972 AJP article "The Copenhagen Interpretation". Mentioning Plato's notion of absolute ideas, he suggested [MM&QM p.76] that perhaps: "It is 'convenient' to consider the ideas as existing even outside of the human mind because otherwise it would be difficult to speak of the world before human ideas have existed." That answer is in line with the science-based conclusion in Chap. 11, that the physically described reality represented by the quantum mechanical state of the universe is most rationally understood as an idea in a universal mind, of which our human minds are tiny partially isolated parts.

A reviewer of this book emphasized the relevance of a 2012 paper of Schurger et al. [21] in which the occurrence of self-initiated movement events are triggered not by mental interventions but by physical fluctuations of a "leaky statistical accumulater". This model is applied to the case in which the standard Libet task of moving a thumb at an unspecified time is modified by an overriding instruction to "immediately" perform the specified action. The new action turns out to be preceded by a more extended in time RP (readiness potential) than when a less urgent, more relaxed, demand is issued.

The empirically observed temporal evolution is in very good agreement with their three-parameter fit, which they interpret as evidence in favor of their purely physicalist model of Libet-type data. But it is not clear that it is evidence against the validity of orthodox mind-dependent vN quantum mechanics. Indeed, the need for a template for a very sharply defined time of action would surely demand longer to prepare, than one for a more loosely defined time of action: more precision with the same tool (the same brain) would require more preparation time. It would certainly be a very major result if the Schurger data were to be actually incompatible the orthodox von Neumann-Copenhagen interpretation. So the Schurger result cannot be an actual problem for the orthodox mind-dependent dynamics being described in this book.

Chapter 11
The Fundamentally Mental Character of Reality

The realistically construed orthodox quantum mechanics described in this book has three components: (1), A physically described universe represented by an evolving quantum mechanical state; (2), An ordered sequence of probing questions that arise in the minds of observers; and (3), A "nature" that chooses and implements—in concordance with Born's statistical rule—psycho-physical responses to the probing questions posed by observers.

The minds of observers, being possessors of thoughts, ideas, and feelings, are "mental" in character, while the quantum state of the universe, being a generalization of the classical state of the universe, is often assumed to have the same ontological character or status as its classical analog.

The classical state of the universe, according to Isaac Newton, is composed of "solid, massy, hard, impenetrable, movable particles". The classical state is thus said to be "matter-like" in character, not "mind-like". It is the carrier of enduring conserved properties such as energy and momentum.

However, the *quantum mechanical* counterpart of the material classical state of the universe represents mere *potentialities* for future psycho-physical happenings. These potentialities are *images of what the future perceptions might be. The state that carries them, like the potentialities they carry is evanescent: it is beset by quantum jumps that are linked to mental events.* Hence the quantum state is more like "an idea" about something, which rapidly changes like an idea does, when new information becomes available, than like a material substance of classical mechanics that tends to endure.

The foregoing summary leads to the conclusion that, in terms of its behavior, the ontological character of quantum reality is more mind-like than matter-like.

That conclusion is far from new. It has been explicitly proclaimed by many distinguished quantum physicists of the past, as the quotations assembled below make clear.

I regard consciousness as fundamental. I regard matter as derivative from consciousness. (Max Planck quoted in the Observer, 25 January 1931.)

"The universe is of the nature of a thought or sensation in a universal Mind." "To put the conclusion crudely – the stuff of the world is mind-stuff". "It is difficult for the

© Springer International Publishing AG 2017
H.P. Stapp, *Quantum Theory and Free Will*,
DOI 10.1007/978-3-319-58301-3_11

matter-of-fact physicist to accept the view that the substratum of everything is of mental character. But no one can deny that mind is the first and most direct thing in our experience, and all else is remote inference – inference either intuitive or deliberate."

(Sir Arthur Eddington, 1928, The Nature of the Physical World, Chap. 13):

In 1961 Erwin Schrödinger wrote:

… it comes naturally to the simple man of today to think of a dualistic relationship between mind and matter as an extremely obvious idea. … But a more careful consideration should make us less ready to admit this interaction of events in two spheres—if they really are different spheres; for the … causal determination of matter by mind …would necessarily have to disrupt the autonomy of material events, while the …causal influence on mind of bodies or their equivalent, for example light…is absolutely unintelligible to us; in short, we simply cannot see how material events can be transformed into sensation or thought, however many text-books, in defiance of Du Bois Raymond, go on talking nonsense on the subject.

These shortcomings can hardly be avoided except by abandoning dualism. This has been proposed often enough, and it is odd that it has usually been done on a materialistic basis. ….But it strikes me that …surrender of the notion of the real external world, alien as it seems to everyday thinking, is absolutely essential.

….If we decide to have only one world, it has got to be the psychic one, since that exists anyway (cogitate—est). And to suppose that there is interaction between the two spheres involves something of a magical ghostly sort; or rather the supposition itself makes them into a single thing. (Schrödinger, My View of the World, pp. 61–63}

Einstein arrives at essentially the same conclusion (of the mental character of the reality implicit in standard quantum theory) when he complains that: "What I dislike about this kind of argumentation is the basic positivistic attitude, which from my point of view is untenable, and which seems to me to come to the same thing as Berkeley's esse est percipi." [To be is to be perceived.] (Albert Einstein: Philosopher-Physicist, Schilpp, p. 669)

Einstein also says that: "What does not satisfy me, from the standpoint of principle, is its attitude toward what seems to me to be the programmatic aim of all physics: the complete description of any (individual) real situation (as it supposedly exists apart from any act of observation or substantiation)." (ibid. p.667)

But Einstein is tacitly demanding concordance with a failed materialistic classical dynamics that is unable to account for the empirical facts. Physics has now advanced to a form that seeks to account only for those aspects of reality that are associated with acts of observation or substantiation. And, as regards the classically supposed irrelevance to reality of our acts of observation or substantiation, William James opined: "It is to my mind quite inconceivable that consciousness should have *nothing to do* with a matter it so closely attends." [20].

What human consciousness does, according to ontologically construed orthodox QM, is to initiate, by its choice of a probing action, a response on the part of nature that actualizes some aspect of reality that was, until then, merely a potentiality. Thus our conscious efforts become causal players in the game of converting potentialities to actualities, and thereby influencing reality.

According to realistically interpreted orthodox QM, we are not the helpless witnesses that classical mechanics claimed us to be, but are, instead, causally effective agents in the creation of an evolving reality. Thus as Bohr repeatedly, and rightfully, reminded us: "In the drama of existence we are ourselves both actors and spectators". That was Heisenberg's seminal 1925 discovery, which constitutes the foundation of our quantum mechanical understanding of the nature of things!

Universal Mind

All that we human beings really know exist are our own mental experiences. But we are relatively recent newcomers to the world revealed by astronomical and archeological observation. Hence there is good reason to believe that there exists, in addition to these evanescent human mental elements, a more enduring reality within which our mental aspects are embedded, or from which they emerge. Thus we can ask: What is the nature or character of this more enduring reality?

The basic message of quantum mechanics is that this underlying reality has, on the basis of its behavior, the same ontological character as the mental realities embedded within it—not the character of the Newtonian-type matter that was postulated to exist by classical mechanics. The underlying reality in quantum mechanics has the ontological character of human thoughts, ideas, and feelings, not the character of solid particles. Thus all of reality is made of one single kind of stuff, and there is no logical problem of the kind that plagues classical physics. Classical mechanics bans the minds of the observer from the matter-based dynamics, whereas quantum mechanics bans Newtonian-type matter from the basic dynamics, and makes mind basic.

Macro-non-locality

The complete lack of micro-level conditions in the Appendix 1 proof of the logical need for essentially instantaneous long-distance information transfer is fatal to theories, or to philosophical positions, that claim that quantum mechanics pertains only to microscopic properties, and hence that the principles of relativistic classical physics work just fine in the realm of large visible properties. The Appendix-1 derivation of quantum non-locality under exclusively macroscopically specified conditions directly refutes that claim.

The failure of many workers on the mind-brain or mind-matter problem to take into account this profound impact of the quantum mechanical character of reality *within the strictly macroscopic realm* has been the source of a widespread pernicious belief that quantum mechanics has little or nothing to do with the big questions of the basic nature of world, and of ourselves. That belief inspires the related notion that the consideration of quantum effects can be relegated to specialists who are interested in the atomic minutiae, while thinkers concerned with the big human issues can pursue their thinking (apart from the intrusion of the quantum elements of random chance) within the simpler framework of classical physics, which excludes our minds from the causal dynamics. But, according to quantum mechanics, the inclusion of the effects of our mental intentions upon the macroscopic behavior of our brains and bodies is absolutely essential to a correct understanding of the dynamical role of our human minds in workings of nature, and hence to a valid self image.

Some quantum physicists have dreamed up non-orthodox ways of trying to capture the quantum aspects of nature, while leaving our minds out of the dynamics. But such theories are necessarily incomplete, compared to the orthodox theory, because they cannot describe the dependence of our physical behavior upon our mental intentions, which are left completely out of the dynamics. A theory that can explain neither the empirical data of atomic physics nor the ubiquitous experienced effects of our intentional known-to-be-real thoughts may be simpler than orthodox quantum mechanics, but is fundamentally deficient, because it is incumbent upon it to explain the data of atomic physics that baffled physicists from the 1913 quasi-classical Bohr model of the atom, until the 1925 introduction by Heisenberg of the non-trivial causal effects of our observational actions.

Chapter 12
Paranormal Phenomena

Quantum theory was originally formulated as a non-relativistic theory. It was assumed that there was a "preferred" coordinate frame that labeled each point in the 4D space-time by three spatial coordinates x, y, and z, plus a time variable t. The quantum state of the full physical universe was defined at each time t, and it evolved from an earlier time t_1 to a later time t_2 by the action of well defined (unitary) operator $U(t_2, t_1)$.

The time t, as defined by this preferred coordinate frame, played a special role: It defined, for each possible value of t, a 3D surface that constituted a possible "present instant now" that separates a t-dependent past from a t-dependant future. The evolving history of the universe was tied to this ever-increasing value of the time t.

The key idea Einstein's theory of relativity was that there should be no dependence of the dynamical rules that generate the evolution of the physical universe on such a preferred frame. The key idea of RQFT was the idea of Tomonaga and Schwinger about how to evade all dynamical dependence on any such preferred frame. They introduced, in place of the flat constant-time 3D surfaces t = constant, the notion of space-like surfaces σ. Every pair of different points on such a surface are connected by a space-like vector.

There is a well-ordered discrete sequence of such surfaces σ(n) such that the integers n label the well-ordered sequence of instants "now" along which the famous "collapses", or "reductions" of the quantum state of the universe occur. The surfaces σ(n) for integer n are temporally ordered in the sense that for some 3D location (x,y,z) the time t(x,y,z,n + 1) is greater than t(x,y,z,n), and for every (x,y,z) the time t(x,y,z,n + 1) is greater than or equal to t(x,y,z,n). And there is a unitary operator U(n'.n) that transforms the state just after the collapse at σ(n) to the state just before the collapse at σ(n').

The unitary U(n',n) is the natural extension of the non-relative U(t',t) from the case of the flat constant-time surfaces t = t(n) surfaces to the case of the non-flat space-like surfaces σ(n). Tomonaga and Schwinger introduce the so-called interaction representation, which has the effect of causing U(t',t) to be the effect of only

© Springer International Publishing AG 2017
H.P. Stapp, *Quantum Theory and Free Will*,
DOI 10.1007/978-3-319-58301-3_12

the "interaction" part of the (Hamiltonian) operator that generates the temporal evolution. This part is local, and hence U(t',t) expresses the local character of the underlying dynamics. U(n',n) inherits this feature.

This all may sound rather complicated, but it merely allows the ideas of von Neumann's detailed theory of measurement, involving the input of an observer's "free choice" of probing action followed by nature's global physical response, to be carried over essentially unchanged to RQFT, with the earlier role of the constant-time 3D collapse surfaces played now by the **frame independent** surfaces σ(n)!

The various paranormal phenomena to be considered here involve a stimulus chosen by a quantum random number generator QRNG and then applied to a living subject, with a physiological response by the subject detected significantly **before** the action of QRNG, and dependent upon the yes-or-no choice generated by the QRNG. Such behavior contradicts the local forward-in-time property of orthodox RQFT.

I do not try to judge validity of the claims of anyone who claims to have found such results, but merely note, first, how such results could be caused by a biasing of the Born Rule in favor of outcomes positively valued by a human being, or favorable to the welfare of some other kind of biological entity. A quantum-like theory modified only in this way, but preserving the rest of the von Neumann described mathematical psycho-physical formalism, I call a "quasi-quantum mechanical theory"!

I shall also describe a very feasible critical experiment that could support, or rule out, this proposed explanation.

The experiment I shall consider is basically the same as the famed "erotic picture" experiment reported by Cornell psychologist Daryl J. Bem. In this experiment a human subject is seated in front of two opaque screens behind each of which lies a computer window. The subject told that behind one of the two screens will soon appear an interesting picture, whereas behind the other will appear only an uninteresting blank wall, and that (s)he should now choose eventually to look behind the screen which (s)he *"feels"* or **"guesses"** the interesting picture will soon appear!

Shortly *after* his or *her choice of the screen to look behind is made,* and securely recorded, a 50–50 random number generator, QRNG1, chooses the screen behind which the interesting picture, not a blank wall, will appear, and another 50–50 random number generator, QRNG2, decides, in the picture case, whether the appearing picture will be erotic or non-erotic. Orthodox QM predicts that in the subset of instances in which either an erotic or non-erotic picture appears the subject will, in ∼50% of the instances, choose to look behind the screen behind which the erotic picture will later appear. But Bem reports that, roughly 53% of the time, the subject chooses, at the outset of the experimental trial to look behind the screen where QNRN2 will **only later** choose to place the erotic picture!

This result seems to demand either clairvoyance, in which the subject's sub-conscious somehow knows in advance what QRNG2 will do, or psychokinesis, in which the observer's mental aspect somehow influences the physical mechanism (of picture placement) tied to QRNG2's random choice. Another possibility is a biasing of nature's Born-Rule-governed choice, at the conclusion of experimental trial, of what the subject will eventually perceive. Such a biasing of nature's final choice of what the observer perceives at the conclusion of the trial could also account for the Bem-reported $\sim 3\%$ seeming aberration of the choice that the subject made at the beginning of the trial.

And there is also the consideration that careful attention must be paid to the important difference between the "actual past" and the "historical past" discussed in Chap. 8, and to the associated profound truth proclaimed by Stephen Hawking and Leonard Mlodinow: "We create history by our observations, history does not create us". Our reconstructed history is based on our *presently existing* records of the past, and these preserved records lack, by virtue of the collapse event, the records that record the processes leading to the possibilities that were ruled out by a collapse event and hence do not survive that collapse event. So the history constructed from the records that exist after a quantum collapse event differs from the history constructed from the records that existed just before that collapse event. History is thus effectively revised not just because of human efforts to re-write it (e.g., for political reasons), but also because of nature's corruption of the records of what was going on before the collapse!

In the Bem experiment under consideration here, if none of the above-mentioned effects are operative, then the fact that nature chooses 50% of the time to exclude all records of the potentialities that were not actualized would mean that in the reconstructed history there never was a potentiality for what did not happen to happen. It would seem like nature's actions **caused** the things that did not happen to not happen: It would seem that, within the orthodox theory, there must be retrocausation!

I believe that there is a clearly favorite option, which I call SPK, or special psycho-kinesis. The proposal to violate the Born Rule is tantamount to abandoning quantum theory, and thus should be put on a far back burner. But SPK respects the Born Rule, and might almost be considered a natural part of orthodox QM.

SPK is a conceivable feature of nature that would allow a conscious agent's mental effort, or intent, to influence a physically described processes lying in the forward light cone of the agent's body, which is a region that includes the agent's body. That one's mind can influence one's future actions is a center-piece of orthodox QM, and much of the paranormal could be explained if this power of mind were to extend beyond the perceiving subject's body. Thus in the Bem experiment the "adventurous" aspects of the subject's mind would be allowed to influence the physical mechanisms that control the placement of the erotic pictures, and influence it in a way that will advance what the subject values or intends. Thus, in accord with "The (reputed) power of positive thinking", (and also remarks of William James) the agent would hold in mind the perceived desiderata, and "nature" would then tend to bring it into being. Such an arrangement would allow our minds to effectively do what standard quantum mechanics says they are doing, together with certain aspects of the paranormal, instead of creating merely:"The illusion of conscious will"

Chapter 13
Conclusions

Scientists discovered in the twentieth century that our Minds Matter: that we are not the pre-programmed mechanical automatons that classical Newtonian physics had, for two hundred years, proclaimed us to be. They explained, moreover, **how** our minds matter: how such intangible things as our mental intentions and consciously felt values influence the behavior of our material bodies. They showed why our mental selves are not, as had been widely believed for two centuries, and are still believed by physicalist philosophies, mere passive witnesses to an inexorable sequence of material events that lie completely beyond the capacity of our thoughts, ideas, and feelings to affect in any way. But quantum mechanics entails that we are not mere material robots deluded by "the illusion of conscious will". Our conscious will is rather, by the means explicitly spelled out in realistically interpreted orthodox von Neumann Quantum theory, able to substantially affect our individual physical lives in ways causally driven by our consciously felt values.

According to the orthodox formulation of quantum mechanics, the basic dynamical process involves a "free choice" on the part of an observer of what perceivable property of the observed system is to be probed, or inquired about. Here the word "free" in "free choice" stipulates that this choice is not fully determined by the material aspects of reality alone, but is influenced by an input from the mind of the observer. This shift in the basic dynamical structure of nature elevates our conscious mental aspects from causally inert by-products of physical brain activity to active participants in the unfolding of a dynamically integrated psycho-physical reality.

This revision of the mind-brain dynamics eliminates the seeming absurdity of a consciousness that exists but can make no difference in what happens. Such a causal inertness is in direct conflict with the ubiquitous evidence of everyday life, which strongly indicates that one's mental intentions normally influence one's bodily behavior in essentially the intended way.

The quantum understanding of the world, involving the causal efficacy of our minds, while incompatible with the prejudices of the "classically scientifically educated" elite, is completely in line with the deeper experience-based intuitive idea

© Springer International Publishing AG 2017
H.P. Stapp, *Quantum Theory and Free Will*,
DOI 10.1007/978-3-319-58301-3_13

of each of us that our mental intentional efforts often influence our physical actions. That experience-based belief is the rational foundation of our meaningful creative lives, and of the societies that we have created to house this intuitive idea of what we are.

The "big" problems of: (1), the connection between mind and brain; (2), "free will"; (3), faster-than-light action-at-a-distance; (4), apparent retrocausal actions; and (5), a rational foundation for morality, have all been addressed in this book within the Orthodox Quantum Mechanical conception of reality, upgraded by the work of Tomonaga and Schwinger to the relativistic form provided by RQFT. That orthodox theory accounts for, in addition to all the new data, also all of the successes of the prior physical theory, classical mechanics, while eliminating its major liabilities, which include: (1), its incompatibility with the findings of atomic physics; (2), its incompatibility with the faster-than-light aspect of nature proved in Appendix 1; and (3), its incompatibility with a belief that is essential to the successful living of our lives, namely the idea that how we physically act is not completely determined either before we were born or by the prior physical reality combined with pure random chance. The orthodox theory rationally explains how our mental intentions, per se, can tend to make our physical actions conform to our value-based mental intentions. That theory thus revokes the classical idea that we are essentially mechanical cogs in a clock-like universe, lacking any capacity to initiate, by mental effort, actions that can aid our survival, advance our values, or improve the world for others.

The simplistic classical conception of reality has thus been converted by a major advance in science to a radically revised image of the cosmos and our place within it. This quantum mechanical conception provides a rationally coherent science-based foundation for human lives suffused with purpose and meaning. A person's mind acts first to construct, from the clues transmitted by sense organs to that observer's brain, a conception of the physically described reality in which the person is locally embedded. The person's mind then directs, in mentally intended ways (via mental efforts that exploit the quantum dynamical laws), that person's bodily actions. Quantum mechanics thereby provides a rational science-based escape from the philosophical, metaphysical, moral, and explanatory dead ends that are the rational consequences of the prevailing entrenched and stoutly defended in practice—although known to be basically false in principle—classical materialistic conception of the world and our place within it.

Acknowledgements The form and content of this book were significantly influenced by advice and suggestions from Ed Kelly, Robert Bennin, Cheryl Sonk, Maria Syldona, Gay Bradshaw, Henry P. Stapp IV, Brian Wachter, Menas Kafatos, and Stan Klein. I warmly thank them all for their help and support.

Supplemental Readings

Bohr N (1935) Can quantum mechanical description of physical be considered complete? Phys Rev 48:696–702

Bohr N (1958) Atomic physics and human knowledge. Wiley, New York

Bohr N (1963) Essays 1958/1962 on atomic physics and human knowledge. Wiley, New York

Einstein A, Podolsky B, Rosen N (1935) Can quantum mechanical description of physical be considered complete? Phys Rev 47:777–780

Einstein A (1951) Remarks to the essays appearing in this collected volume. In: Schilpp PA (ed), Albert Einstein: philosopher-physicist. Tudor, New York

Libet B (1985) Unconscious cerebral initiative and the role of conscious will in voluntary action. Behav Brain Sci 8, 529–566

Libet B (2003) Cerebral physiology of conscious experience: experimental studies. In: Osaka N (ed.), Neural basis of consciousness. Advances in consciousness research series, vol 49, John Benjamins, Amsterdam & New York

James W (1892) Psychology: the briefer course. In: William James: writings 1879–1899. Library of America, New York

James W (1911) Some Problems in Philosophy. In: William James; writings 1902–1910. Library of America, New York

Rosenfeld L (1967) In Wheeler and Zurek (eds.), Quantum theory and measurement. Princeton University Press, Princeton N.J

Schwartz J, Stapp H, Beauregard M (2005) Quantum physics in neuroscience and psychology: a neuro-physical model of mind/brain interaction. Rhol Trans R Soc B 360 (1458) 1308–27. (http://www-physics.lbl.gov/~stapp/PTRS.pdf)

Schwinger J (1951) Theory of quantized fields I. Phys Rev 82:914–27.

Stapp HP (2002) The basis problem in many-worlds theories, Can J Phys 80:1043–1052

Stapp HP (2004a) Mind, matter, and quantum mechanics, 2nd edn. Springer, Berlin, Heidelberg, New York

Stapp HP (2004b) A Bell-type theorem without hidden variables. Am J Phys 72:30–33

Stapp HP (2005) Quantum interactive dualism: an alternative to materialism. J Conscious Stud 12(11): 43–58

Stapp HP (2006a) Quantum approaches to consciousness. In: Moskovitch M, Zelago P (eds), Cambridge Handbook of Consciousness, Cambridge: Cambridge U. P.

Stapp HP (2006b) Quantum mechanical theories of consciousness. In Velmans M, Schneider S (eds), Blackwell companion to consciousness. Blackwell Publishers, Oxford

Stapp HP (2006c) The quest for consciousness: a quantum neurobiological approach, see http://www-physics.lbl.gov/~stapp/Quest.pdf

Stapp HP (2006d) Mindful universe, see (http://www-physics.lbl.gov/~stapp/MU.pdf)

Tomonaga S (1946) On a relativistically invariant formulation of the quantum theory of wave fields. Prog Theor Phys 1: 27–42

Von Neumann J (1955/1932) Mathematical foundations of quantum mechanics. Princeton University Press, Princeton. (Translated by Robert T. Beyer from the 1932 German original, MathematicheGrundlagen der Quantummechanik. Berlin: J. Springer)

Appendix A
Proof that Information Must So
be Transferred Faster Than Li

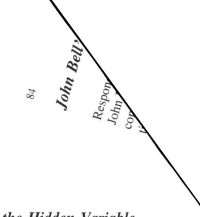

Strengthening Bell's Theorem by Removing the Hidden-Variable Assumption

In the context of correlation experiments involving pairs of experiments performed at essentially the same time in very far-apart experimental regions Einstein famously said [22]:

> But on one supposition we should in my opinion hold absolutely fast: 'The real factual situation of the system S2 is independent of what is done with system S1 which is spatially separated from the former.'

This demand is incompatible with the basic ideas of standard (Copenhagen/Orthodox) quantum mechanics, which makes two relevant claims:

(1) Experimenters in the two labs make "local free choices" that determine which experiments will be performed in their respective labs. These choices are "free" in the sense of not being pre-determined by the prior history of the physically described aspects of the universe, and they are "localized" in the sense that the physical effects of these free choices are inserted into the physically described aspects of the universe only within the laboratory, and during the time interval, in which the associated experiment is being performed.

(2) These choices of "what is done with the system" being measured in one lab can (due to a measurement-induced global collapse of the quantum state) influence the outcome of the experiment performed at very close to the same time in the very faraway lab.

This influence of 'what is done' with the system being measured in one region upon the outcome appearing at very nearly the same time in a very faraway lab was called "spooky action at a distance" by Einstein, and was rejected by him as a possible feature of "reality".

, *Quasi-classical Statistical Theory*

ꞏng to the seeming existence in the quantum world of "spooky actions",
ell [23] proposed a possible alternative to the standard approach that might
ꞏeivably be able to reconcile quantum spookiness with "reality". This alterna-
ꞏe approach rests on the fact that quantum mechanics is a statistical theory. We
already have in physics a statistical theory called "classical statistical mechanics".
In that theory the statistical state of a system is expressed as a *sum* of terms, each of
which is a possible *real physical state* λ of the system multiplied by a probability
factor.

Bell conjectured that quantum mechanics, being a statistical theory, might have
the same kind of structure. Such a structure would satisfy the desired properties of
"locality" and "reality" (local realism) if, for each real physical state λ in this sum,
the relationships between the chosen measurements in the two regions and the
appearing outcomes are expressed as product of two factors, with each factor
depending upon the measurement and outcome in just one of the two regions. The
question is then whether the statistical properties of such a statistical ensemble can
be consistent with the statistical predictions of quantum mechanics.

Bell and his associates proved that the answer is No! They considered, for
example, the empirical situation that physicists describe by saying that two spin-1/2
particles are created in the so-called spin-singlet state, and then travel to the two
far-apart but nearly simultaneous experimental regions. The experimenter in each
region freely chooses and performs one of the two alternative possible experiments
available to him. Bell et. al. then prove that the predictions of quantum mechanics
cannot be satisfied if the base states λ satisfy the "factorization property" demand of
"local realism". A theory satisfying this demand is called a "local hidden-variable
theory" because the asserted underlying "reality" is described by variables that
cannot be directly apprehended.

Two Problems with Bell's Theorems

Bell-type theorems, if *considered as proofs of the logical need for spooky actions in
a theory that entails the predictions of quantum mechanics*, have two problems. The
first is that the theorems postulate a "reality" structure basically identical to that of
classical statistical mechanics. Bell's theorems then show that imposing "locality"
(factorizability for each fixed λ) *within this classical-type reality structure* is
incompatible with some predictions of quantum mechanics. But that result can be
regarded as merely added confirmation of the fact that quantum mechanics is
logically incompatible with the conceptual structure of classical mechanics. Simply
shifting to a classically conceived "statistical" level does not eliminate the essential
conceptual dependence on the known-to-be-false concepts of classical physics.

The second problem is that the condition of "local realism" is implemented by a "factorization" property, described above, that goes far beyond Einstein's demand for no spookiness. In addition to the non-dependence of outcomes in a region upon *"what is done"* in the faraway region "local realism" entails also what Shimony calls "outcome independence". That condition goes significantly beyond what Einstein demanded, which is merely a non-dependence of the factual reality (occurring outcome) in one region on *the choice of experiment performed in the faraway region.* "Outcome independence" demands that the outcome in each region be independent also of the *outcome* in the other region.

That property, *"outcome independence"*, is not something that one wants to postulate if a resulting incompatibility with predictions of quantum mechanics is supposed to entail the existence of spooky actions at a distance!

That unwanted independence assumption is not a just a minor fine point. Consider the simple example of two billiard balls, one black, one white, shot out in opposite directions to two far-apart labs. This physical example allows—given the initial symmetrical physical state—the outcome in one region to be correlated with the *"outcome"* appearing in the other region, without any hint of any spooky action at a distance": a "black" ball in one region entails a "white" ball in the other, and vice versa, without any spooky action. Hence Bell's theorems do not address—or claim to address—the key question of the compatibility of Einstein's demand for no spookiness with the predictions of (relativistic) quantum mechanics. Bell's theorems are based on the stronger assumption of local hidden variables.

Bell's theorems (regarded as proofs of the need for spooky actions) are thus deficient in two ways: they bring in from classical (statistical) mechanics an alien-to-quantum-mechanics idea of "reality"; and they assume, in the process of proving a contradiction, a certain property of "outcome independence" that can lead to a violation of quantum predictions without entailing the lack of spookiness that Einstein demanded.

The question thus arises whether the need for spooky interactions can be proved simply from the validity of some empirically well validated predictions of standard quantum mechanics, without introducing Bell's essentially classical "hidden variables"? The answer is "Yes"!

The Proof

The following proof of the need for "spooky actions" places no conditions at all on any underlying process or reality, beyond the macroscopic predictions of quantum mechanics: it deals exclusively with connections between macroscopic measurable properties. This change is achieved by taking Bell's parameter λ to label, now, the different experiments in a very large set of simultaneously performed similar experiments, rather than the different possible basic microscopic states λ of the statistical ensembles. The ontology thereby becomes essentially different, though the mathematics is similar. The macroscopic experimental arrangements are the ones already described above.

In the design of this experiment the physicists are imagining that a certain initial macroscopic preparation procedure will produce a pair of tiny invisible (spin 1/2) particles in what is called the singlet state. These two particles are imagined to fly out in opposite directions to two faraway experimental regions. Each of these experimental regions contains a Stern-Gerlach device that has a directed preferred axis that is perpendicular to the incoming beam. Two detection devices are placed to detect particles deflected either along this preferred axis, or in the opposite direction. Each of these two devices will produce a visible signal (or an auditory click) if the imagined invisible particle reaches it.

The location of the individual detector is specified by the angle ϕ of the directed preferred axis such that a displacement along that particular direction locates the detector. Clearly, the two detectors in the same experimental region will then be specified by two angles ϕ that differ by 180°. For example, if one detector is displaced "up" ($\phi = 90°$) then the other is displaced "down" ($\phi = -90°$). The angle $\phi = 0$ labels in both regions a common deflection to the right: e.g., along the positive x axis in the usual x-y plane.

Under these macroscopic experimental conditions, quantum theory predicts that, if the detectors are 100% efficient, and if, moreover, the geometry is perfectly arranged, then for each created pair of particles—which are moving in opposite directions to the two different regions—exactly one of the two detectors in each region will produce a signal (i.e., "fire"). The key prediction of quantum theory for this experimental setup is that the fraction F of the particle pairs for which the detectors that fire in the first and second regions are located at angles ϕ_1 and ϕ_2, respectively, is given by the formula

$$F = (1 - Cosine(\phi_1 - \phi_2))/4.$$

In the experiment under consideration there are two alternative possible experiments in the left-hand lab, and two alternative possible experiments in the right-hand lab, making $2 \times 2 = 4$ alternative possible pairs of experiments. For each single experiment (on one side) there are two detectors, and hence two angles ϕ. Thus there are altogether $4 \times 2 \times 2 = 16$ F's.

I take the large set of similar experiment to have 1000 experiments. Then the fractions F of 1000 are entered into the 16 associated boxes of the following diagram.

In Fig. A1, the first and second *rows* correspond to the two detectors in the *first* possible set-up in the left-hand region. The third and fourth rows correspond to the two detectors in the *second* possible set-up in the left-hand region. The four *columns* correspond in the analogous way to the detectors in the right-hand region. The arrows on the periphery show the directions of the displacements of the detectors associated with the corresponding row or column.

For example, in the top-left 2-by-2 box if the locations of the two detectors (one in each region) that fire together are both specified by the same angle, $\phi_1 = \phi_2$,

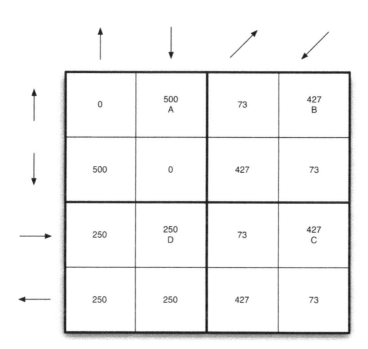

Fig. A1 Matrix of transition probabilities described in the text

then, because *Cosine 0 = 1*, each specified pair of detectors will *never* both fire together: if one of these two specified detectors fires, then the other will not fire. If ϕ_1 is some fixed angle and ϕ_2 differs from it by 180° then, because *Cosine 180° = −1*, these two specified detectors will, under the ideal measurement conditions, fire together for *half* of the created pairs. If ϕ_1 is some fixed angle and ϕ_2 differs from it by 90° then these two specified detectors will fire together for $1/4$ of the pairs. If ϕ_1 is some fixed angle and ϕ_2 differs from it by 45° then these two specified detectors will fire together, in a long run, for *close to 7.3%* of the pairs. If ϕ_1 is some fixed angle and ϕ_2 differs from it by 135° then these two specified detectors will fire together, in a long run, for *close to 42.7%* of the created pairs.

I have listed these particular predictions because they are assumed to be valid in the following proof of the need for near-instantaneous transfer of information between the two far-apart, but nearly simultaneous, experimental space-time regions. These particular predictions have been massively confirmed empirically.

The second assumption is "localized free choices". The point here is that physical theories make predictions about experiments performed by experimenters with devices that detect or measure properties of the systems whose properties are being probed by these devices. The theory entails that the various settings of the devices will correspond to probe-associated properties of the system being probed.

Of course, in an actual situation these specified parts of the experimental setup are all parts of a universe that includes also the experimenter and whatever the experimenter uses to actually fix the experimental settings. Such a "choosing" part of the universe *could*, however, *conceivably* causally affect not only to the setting of the associated measuring device but, say, via the distant past, also other aspects of the experiment. Those unsuspected linkages via the past could then be responsible for systematic correlations between the empirical conditions in the two regions— correlations that are empirically dependent on which experiments are chosen and performed but are empirically independent of <u>how</u> the experimental setups are chosen.

In view of the limitless number of ways one could arrange to have the experimental setup specified, and the empirically verified fact that the predictions are found to be valid independently of <u>how</u> the setup is chosen, it is reasonable to assume that the choices of the experimental setups can be arranged so that they are not systematically connected to the specified empirical aspects of the experiment except via these choices of the experimental setup. This is the assumption of "localized free choices." It is needed to rule out the (remote) possibility that the choice of the setup is significantly and systematically entering the dynamics in some way other than as just the localized fixing of the experimental setup.

Suppose, then, that we have the two far-apart experimental regions, and in each region an experimenter who can freely choose one or the other of two alternative possible experimental set-ups. Suppose we have, in a certain region called the source region, a certain mechanical procedure to which we give the name "creation of N individual experimental instances, where N is a large number, say a thousand. At an appropriate later time the experimenters in the two regions make and implement their "localized free choices" pertaining to which of the two alternative possible experiments will be set up in their respective experimental regions. At a slightly later time each of the two experimenters looks at and sees, in each of the N individual instances, which one of his two detection devices has fired, and then records the angle ϕ that labels that detector, thereby recording the outcome that occurs in that individual instance.

There are altogether two times two, or four, alternative possible experimental setups. Figure A1 gives, for each of these four alternative possible setups, the number of individual instances, from the full set of 1000, that produce firings in the pair of detectors located at the pair of angles ϕ specified along the left-hand and top boundaries of the full diagram. For example, the four little boxes in the first two rows and the first two columns correspond to the case in which the experimenter in the left-hand region sets his two detectors at "up" ($\phi_1 = 90°$) and "down" ($\phi_1 = -90°$), while the experimenter in the right-hand region sets his two detectors also at "up" ($\phi_2 = 90°$) and "down" ($\phi_2 = -90°$). In this case the expected distribution (modulo fluctuations) of the thousand instances is 500 in the box in which $\phi_1 = 90°$ and $\phi_2 = -90°$ and the other 500 in the box in which $\phi_1 = -90°$ and $\phi_2 = 90°$.

The fluctuations become relatively smaller and smaller as N get larger and larger. So I will, for simplicity, ignore them in this discussion and treat the predictions to be exact already for N = 1000.

The two experimental regions are arranged to be essentially simultaneous, *very* far apart, and very tiny relative to their separation. These two regions will be called the "left" and "right" regions.

The "no-essentially-instantaneous-transfer of information about localized free choices" assumption made here is that, no matter which experiment is performed in a region, the outcome appearing there is independent of which experiment is freely chosen and performed in the faraway region. This means, for example, that if the experiment on the right is changed from the case represented by the left-hand two columns to the case represented by the right-hand two columns, then the *particular set* of 500 instances—from the full set of 1000—that are represented by the 500 in the top row second column get shifted into the two boxes of the top row in the second two columns.

More generally, a change in the experiment performed on the right shifts the individual instances—in the set of 1000 individual instanced—horizontally, in the same row; whereas a change in the experiment performed on the left shifts the individual instance vertically. The Fig. A1 then shows how, by a double application of the "no FTL condition", a subset of the set of 500 instances occupying box A gets shifted via box B to box C, which must then contain at least $427 - 73 = 354$ of the original 500 instances in A. However, the applying of the two changes in the other order, via D, demands that the subset of instances in A that can be in C can be no greater than 250. That is a contradiction. Thus one cannot maintain simultaneously both the general rule of no FTL transfers of information and four very basic and empirically confirmed predictions of quantum mechanics.

In more detail the argument then goes as follows. Let the pairs (individual instances) in the ordered sequence of the 1000 created pairs be numbered from 1 to 1000. Suppose that the actually chosen pair of measurements corresponds to the first two rows and the first two columns in the diagram. This is the experiment in which, in each region, the displacements of the two detectors are "up" and "down". Under this condition, quantum theory predicts that for some particular 500-member subset of the full set of 1000 individual instances (created pairs) the outcomes conform to the specifications associated with the little box labeled A. The corresponding 500 member subset of the full set of 1000 positive integers is called Set A. This Set A is a particular subset of 500 integers from set $\{1, 2, ...,1000\}$. The first 4 elements in Set A might be, for example, $\{1, 3, 4, 7\}$.

If the local free choice in the right-hand region had gone the other way, then the prediction of quantum mechanics is that the thousand integers would be distributed in the indicated way among the four little boxes that lie in one of the first two rows and also in one of the *second* two columns, with the integer in each of these four little boxes specifying the number of instances in the subset of the original set of 1000 individual instances that lead to that specified outcome. Each such outcome consists, of course, of a pair of outcomes, one in the left-hand experimental region, and specified by the row, the other in the right-hand experimental region and specified by the column.

If we now add the Locality Condition, then the demand that the macroscopic situation in the left-hand region be undisturbed by the reversal of the localized free choice made by the experimenter in the (faraway) right-hand region means that the set of 500 integers in Set A must be distributed between the two little boxes standing directly to the right of the little box A. Thus the Set B, consisting of the 427 integers in box B, would be a 427 member subset of the 500 integers in Set A.

The above conclusions were based on the supposition that the actual choice of experiment on the left was the option, represented by the top two rows and the leftmost two columns in Fig. A1. However, having changed the choice in the right-hand region to the one that is represented by the *rightmost* two columns—the possibility of which is which is entailed by Einstein's reference to a dependence on "what is done with" the faraway system—we next apply the locality hypothesis to conclude that changing the choice on the left must leave the outcomes on the right undisturbed. That means changing the top two rows to the bottom two rows, leaving the integers that label the particular experiment in the set of 1000 experiments in the same column. This means that the 427 elements in the box B must get distributed among the two boxes that lie directly beneath it. Thus box C must include at least $427 - 73 = 354$ of the 500 integers in Set A.

Repeating the argument, but reversing the order in which the two reversals are made, we conclude, from exactly the same line of reasoning, that box C can contain no more than 250 of the 500 integers box A, Thus the conditions on Set C that arise from the two different possible orderings of the two reversals are contradictory!

A contradiction is thus established between the consequences of the two alternative possible ways of ordering these two reversals of localized free choices. Because, due to the locality hypothesis being examined, no information about the choice made in either region is present in the other region, no information pertaining to the order in which the two experiments are performed is available in either region. Hence nothing pertaining to outcomes can depend upon the relative ordering of these two space-like separated reversals of the two choices.

This argument uses only macroscopic predictions of quantum mechanics—without any conditions on, or mention of, any micro-structure from whence these macroscopic properties come—to demonstrate the logical inconsistency of combining a certain 16 (empirically validated) predictions of quantum mechanics with the locality hypothesis that for each of the two experimental regions there is no faster-than-light transfer to the second region of information about macroscopically localized free choices made in the first.

The Bell's theorem proofs are rightly identified as proofs of the incompatibility of "local realism" with the predictions of quantum mechanics. But "local realism" brings in both alien-to-quantum-theory classical concepts and also an "outcome independence" condition whose inclusion nullifies those theorems as possible proofs of the need for spooky actions at a distance. Both of these features are avoided in the present proof.

As regards Einstein's reality condition, namely that the no-spooky-action condition pertains to the "real factual situation" one must, of course, use the quantum conception of the "real factual situation", not an invalid classical concept. In ontologically construed orthodox quantum mechanics (in the contemporary relativistic quantum field theory version that I use) the "real factual situation" evolves in a way that depends upon the experimenter's free choices and nature's responses to those choices. The no-spooky-action condition is a condition on these choice-dependent real factual situations—namely outcomes observed under the chosen conditions—that is inconsistent with certain basic predictions of the theory. That is what has just been proved. In classical mechanics there are no analogous free choices: the physical past alone uniquely determines the physical present and future.

The Einstein idea of no spooky actions involves comparing two or more situations only one of which can actually occur. This is the kind of condition that occurs in modal logic considerations involving "counterfactuals". But here this modal aspect does not bring in any of the subtleties or uncertainties that plague general modal logic. For in our case the specified condition is a completely well defined and unambiguous (trial) mathematical assumption of the non-dependence of a nearby outcome upon a faraway free choice between two alternative possible probing actions. The proof does not get entangled with the subtle issues that arise in general modal logic. Everything is just as well defined as in ordinary logic.

In this proof there is no assumption of a "hidden variable" of an essentially classical kind lying "behind" the ontologically construed orthodox quantum theory. The phenomena are rationally understandable in terms of an evolving quantum state of the universe that represents "potentialities for experiences" that evolve via a Schrödinger-like equation punctuated by an ordered sequence of psycho-physical events each of which is an observer's personal experience accompanied by a "collapse of the quantum state of the universe" that brings that evolving state into conformity with the observer-initiated experience of that observer.

The bottom line is that, given the validity of some basic macroscopic predictions of quantum mechanics, there is no way that the macroscopic phenomena can conform to the predictions of quantum mechanics without allowing violations of the general notion that the information about the local free choices cannot get essentially instantaneously to faraway regions and affect outcomes appearing in those regions.

Appendix B
Graphical Representation of the Argument

The argument in Appendix A was expressed in words and equations. For many purposes it is useful, for arguments involving an ordered sequence free choices or decisions, to have a graphical representation of the alternative possibilities

The argument in Appendix A is based on statements of the form:

> If measurement M is performed and the outcome is O, then if, instead of M, the measure M' were to be performed, then the outcome would be O'.

Statements of this kind make sense in classical physics. An outcome O of M could, within some theoretical framework, give some information about the state of the world before the measurement M was performed, and this information could entail that O' would occur if M' were to be performed. For example, the outcome of the first experiment could give information about the previously unknown or unspecified velocity of a particle entering the experimental region, and this added information could allow the outcome O' of M' to be predicted. The connection between the two alternative possible situations is a consequence of the conjectured structure of the reality lying behind the observable phenomena. It is therefore a condition on the real existence of that conjectured structure.

The argument in Appendix A involves only macroscopic choices of measurements and outcomes, and a conjectured no-faster-than-light condition. These things are all classically understandable, and the argument can be represented graphically.

Statements of this kind can be definitely true or definitely false in the context of a physical theory that has logically consistent laws that allow the "free choices" between which of several alternative possible experiments is performed to be treated as free variables. Copenhagen and orthodox quantum mechanics are theories of this kind.

Logical reasoning is aided by having a "mechanical" way of checking the truth or falsity of statements. Then all competent users of the logic can agree on the truth or falsity of the propositions.

Robert Griffiths [13] has invented such a "mechanical" procedure for validating reasoning of this kind. It is a graphical procedure. It involves a tree graph that, reading from left to right, has branches that "branch" at branch points into more branches. Some branch points represent the occurrence of events where a choice must be made between two (or more) alternative possible experiments. Other

© Springer International Publishing AG 2017
H.P. Stapp, *Quantum Theory and Free Will*,
DOI 10.1007/978-3-319-58301-3

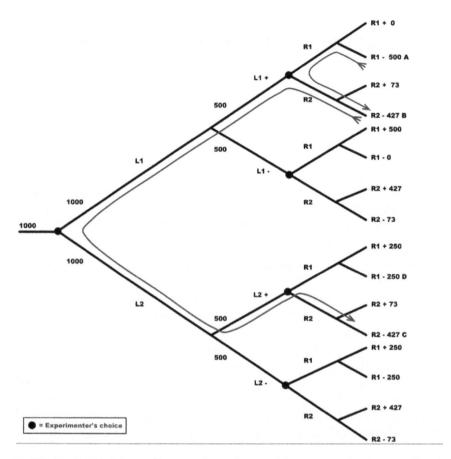

Fig. B1 The Griffiths Diagram Corresponding to the part of the argument given in Appendix A in which the reversal R1 to R2 in region R precedes the reversal L1 to L2 in region L

branch points represent events where some particular outcome of some particular experiment must be chosen (by nature).

If, as in our case, there are two far-apart experimental regions, then the full graphical part that represents the possible events in the later region must be hooked onto each of the branches representing an outcome in the first region, in order for the graph, reading from left to right, to represent, without prejudice stemming from the no-faster-than-light conjecture at issue, the temporal order of the macroscopic events.

Griffiths allows graphs that include branch points corresponding to microscopic (invisible) events, but I exclude all such points and consider only visible events. For the argument in Appendix A explicitly precludes all reference to such imagined events.

Figures B1 and B2 give the graphical representations of the two parts of the argument in Appendix A. The part of the graph that corresponds to the part of the

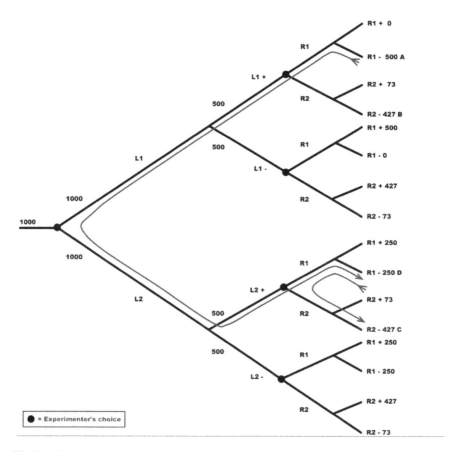

Fig. B2 The Griffiths Diagram Corresponding to the part of the argument given in Appendix A in which the reversal L1 to L2 in region L precedes the reversal R1 to R2 in region R

process labeled L (for left-hand region) stands to the left of the parts labeled R (for right-hand region). The left-to-right ordering in the graph corresponds to increasing time. Thus the L part of the physical process is earlier than the R part.

The argument in Appendix A involves two different orderings of the reversals. So one might consider a second graph with the L-R ordering reversed. But a key requirement of Griffiths' formalism is that a valid argument must be expressed by using only one single graph. So, within Griffiths' theory, the reasoning in Appendix A must be justified by using only one single graph. Consequently, the two parts of the argument must use the same graph. The superposed thinner lines in the two diagrams represent the propositions in the two parts of the argument.

Figure B1 represents the case in which the reversal from experiment R1 to experiment R2 comes first. Keeping track of the 500 elements of Set A under this reversal, which leaves everything in region L unchanged, we see that 427 of the 500 elements a Set A go to Set B. Next comes the reversal L1 to L2 in region L, with the experimenter's choice of R2 in region R left unchanged. We are interested in how many of the 500 elements in set A end up in set C, which corresponds to L2+. These must come from the 427 elements in set B. Because at most 73 of these 427 elements can go to R2+, at least $427 - 73 = 354$ must end up in set C. This is just a diagrammatic representation in the pertinent Griffiths graph of the first half of the argument given in Appendix A.

Figure B2 represents the second half of the argument given in Appendix A, the part in which the first reversal is the reversal of L1 to L2 with the choice in region R of R1 held fixed. Starting again with the 500 elements in set A, but now tracing first back to the experimenter's choice between L1 and L2, and then forward along the other branch, L2, and following the L2+ branch that leads to branch D. Only 250 0f the original 500 instances in Set A end up in D. Then the reversal of R1 to R2 keeping the choice in region L of L2 unchanged allows at most 250 of the elements in Set A to be in Set C. This conclusion conflicts with the conclusion associated with Fig. B1, which was that at least 354 elements of set A are contained in set C. Thus the conclusion deduced in Appendix A by using the common-sense under-standings of the meanings of the words is confirmed within Griffiths' graphical representation of the structure of counterfactual reasoning, restricted now to visible macroscopic events.

A variation of this argument based on experiments of the kind proposed by Julian Hardy has been described in [14, 15, 16, 17, 18, 19]. The argument given above is the one given earlier in [18], here spelled out in greater detail.

Appendix C
Reply to Sam Harris on Free Will

Sam Harris's book "Free Will" is an instructive example of how a spokesman dedicated to being reasonable and rational can have his arguments derailed by a reliance on prejudices and false presuppositions so deep-seated that they block seeing science-based possibilities that lie outside the confines of an outmoded world view that is now known to be incompatible with the empirical facts.

A particular logical error appears repeatedly throughout Harris's book. Early on, he describes the deeds of two psychopaths who have committed some horrible acts. He asserts: "I have to admit that if I were to trade places with one of these men, atom for atom, I would be him: There is no extra part of me that could decide to see the world differently or to resist the impulse to victimize other people."

Harris asserts, here, that there is "no extra part of me" that could decide differently. But that assertion, which he calls an admission, begs the question. What evidence rationally justifies that claim? Clearly it is not empirical evidence. It is, rather, a prejudicial and anti-scientific commitment to the precepts of a known-to-be-false conception of the world called classical mechanics. That older scientific understanding of reality was found during the first decades of the twentieth century to be incompatible with empirical findings, and was replaced during the 1920s, and early 1930s, by an adequate and successful revised understanding called quantum mechanics. This newer theory, in the rationally coherent and mathematically rigorous formulation offered by John von Neumann, features a separation of the world process into (1), a physically described part composed of atoms and closely connected physical fields; (2), some psychologically described parts *lying outside the atom-based part*, and identified as our thinking ego's; and (3), some psycho-physical actions attributed to nature. Within this empirically adequate conception of reality there *is* an extra (non-atom-based) part of a person (his thinking ego) that can resist (successfully, if willed with sufficient intensity) the impulse to victimize other people. Harris's example thus illustrates the fundamental errors that can be caused by identifying honored science with nineteenth century classical mechanics.

Harris goes on to defend "compatibilism", the view that claims both that every physical event is determined by what came before in the physical world and also that we possess "free will". Harris says that "Today the only philosophically

© Springer International Publishing AG 2017
H.P. Stapp, *Quantum Theory and Free Will*,
DOI 10.1007/978-3-319-58301-3

respectable way to endorse free will is to be a compatibilist—because we know that determinism, in every sense relevant to human behavior, is true".

But what Harris claims that "We know" to be true is, according to quantum mechanics, not known to be true.

The final clause "*in every sense relevant to human behavior*" is presumably meant to discount the relevance of quantum mechanical indeterminism, by asserting that quantum indeterminism is not relevant to human behavior—presumably because it washes out at the level of macroscopic brain dynamics But that idea of what the shift to quantum mechanics achieves is grossly deficient. The quantum indeterminism merely opens the door to a complex dynamical process that not only violates determinism (the condition that the physical past determines the future) at the level of human behavior, but allows mental intentions that are not controlled by the physical past to influence human behavior in the intended way. Thus the shift to quantum mechanics opens the door to a causal efficacy of free will that is ruled out by Harris's effective embrace of false nineteenth science.

Appendix D
The Paranormal and the Principle of Sufficient Reason

This book has been an exposition of what I call "Realistically construed orthodox quantum mechanics". That name is intended to mean the conception of reality most naturally concordant with the "pragmatic" interpretation offered by the founders (Heisenberg, Schrödinger, Born, Dirac, Bohr, and Pauli), as mathematically formulated by John von Neumann and subsequently converted during the late 1940s (principally by Tomonaga, Schwinger, and Feynman) into contemporary Relativistic Quantum Field Theory, RQFT.

It might be objected that this orthodox theory is deficient because it does not encompass the widely reported paranormal phenomena. On the other hand, many scientists believe that those phenomena, which, by definition, are incompatible with the generally accepted contemporary physical theory, ought not be considered to be "science" because, by and large, the papers reporting them have not been published in the most prestigious scientific journals, ostensibly because of defects in their methods and procedures.

One exception to that publication criterion is a recent (2012) paper in the Journal of Personality and Social Psychology by Cornell psychologist Daryl J. Bem. However, I show in this appendix how Bem's data, which *appears* to require backward-in-time (retro-causal) effects forbidden by the strictly orthodox RQFT can be explained without any actual backward-in-time action by passing from this "strictly orthodox" theory to a modified "quasi-orthodox" version which, unlike the orthodox theory, enforces "The principle of sufficient reason". This principle asserts that every definite happening must have some definite reason to be what it is, rather than something else. This principle seems, from a general scientific point of view, to be rational and reasonable: it bans the possibility that a definite value of a physical property can just suddenly "pop out of the blue", which is what the strictly orthodox theory effectively demands when it specifies that nature makes strictly random yet perfectly definite choices. The reasonable question is: "What principle or process separates this chosen value from the un-chosen ones, in the (assumed to be single) observed universe?"

It turns out that Bem's data can be accommodated by replacing the "strictly-orthodox" theory by a "quasi orthodox" theory that upholds the principle of sufficient reason. In the quasi-orthodox theory nature's choices are not random, hence lacking a sufficient reason to be what they are, but are assumed to have

© Springer International Publishing AG 2017
H.P. Stapp, *Quantum Theory and Free Will*,
DOI 10.1007/978-3-319-58301-3

sufficient reasons that entail a biasing of nature's choices in favor of choices that advance the personal values of the subjects in these experiments who are posing the questions pertaining to matters of interest to themselves.

In the Bem experiment these pertinent choices occur at the very end of the experimental instance, and hence later in time than another "independent" choice made by the subject-observer at the beginning of the experimental instance. However, a biasing of the later choice biases *records of the past*, due to the difference between the actual and effective past discussed in Chap. 7. Thus, on the basis of the recorded data that survive the collapses, the experimenter will conclude that the earlier choice was biased. That is because the records that survive the final collapse will not include the records residing in the branch of reality that was created at the moment the device made the macroscopic random choice of which picture to present to the subject, but that nature later (biasly) chose not to actualize.

To explain this possibility in more detail, I include (intact) in this appendix a paper that I prepared in 2012 but never submitted to a journal. It shows that the retro-causal actions seemingly found by Bem can be avoided by replacing the "irrational" strictly orthodox dependence of nature's choice of response to the subject's probing actions upon "pure chance" by a meaningful choice that depends on values residing in that person's "ego". That biasing of the statistical weight of a choice made by nature at the *tail end* of the experimental instance effectively biases, via the quantum collapse, the weighting of the "effective" past that precedes that collapse without there being any actual backward-in-time action.

Quasi-Orthodox Quantum Mechanics and the Principle of Sufficient Reason

Abstract The principle of sufficient reason asserts that anything that happens does so for a reason: no definite state of affairs can come into being unless there is a sufficient reason why that particular thing should happen instead of something else. This principle is usually attributed to Leibniz, although the first recorded Western Philosopher to use it was Anaximander of Miletus. The demand that nature be rational, in the sense that it be compatible with the principle of sufficient reason, conflicts with a basic feature of contemporary orthodox physical theory, namely the notion that nature's responses to the probing actions of observers be determined by pure chance, and hence on the basis of absolutely no reason at all. This injection of "irrational" pure chance can be deemed to have no fundamental place in reason-based Western science, and it has been criticized by Einstein, among others. It is argued here that in a world that conforms to the principle of sufficient reason, the usual quantum statistical rules will naturally emerge at the pragmatic level, in cases where the reason behind Nature's choice of response is unknown, but that the usual statistics can become biased in an empirically manifest and apparently retrocausal way when the reason for the choice is empirically identifiable. It is shown here that some recently reported high profile experimental results that violate the principles of contemporary physical theory can be rationally and simply explained if nature's supposedly random choices are sometimes slightly biased in a way that depends upon the emotional valence of the observer-experiences that these choices create.

Introduction

An article recently published by the Cornell psychologist Daryl J. Bem [24] in a distinguished psychology journal has provoked a heated discussion in the New York Times [25]. Among the discussants was Douglas Hofstadter who wrote that: "If any of his claims were true, then all of the bases underlying contemporary science would be toppled, and we would have to rethink everything about the nature of the universe."

It is, I believe, an exaggeration to say that if any of Bem's claims were true then "all of the bases underlying contemporary science would be toppled" and that "we would have to rethink everything about the nature of the universe". In fact, all that is required is a relatively small change in the rules, and one that seems even more reasonable and natural than the usual rules, within the broad general framework of rational Western science. The major part of the required rethinking was done already by the founders of quantum mechanics, and cast in more rigorous form by John von Neumann [26], more than seventy years ago.

According to the ordinary precepts of *classical* mechanics, once the physically described universe is created, it evolves continuously in a deterministic manner that is completely fixed by mathematical laws that depend always and everywhere only on the evolving local values of physically described properties. There are no inputs into the dynamics that go beyond what is specified by those physically described properties. Here *physically described properties* are properties that are specified by assigning mathematical properties to space-time points, or to very tiny regions, independently of whether they are presently being experienced by any biological or other experiencing entity. These properties are thereby distinguished from properties that are described directly in terms of *actually experienced* thoughts, ideas, or feelings. Within that classical mechanical framework of physics the increasing experienced knowledge of human beings and other biological agents enters only as an *output* of the physically described evolution of the universe: experiential aspects of reality that go beyond the purely physical aspects play no role in the algorithmically determined mechanistic evolution of the universe, except perhaps at its birth.

This one-way causation from the physical aspects of nature to the empirical/epistemological/mental aspects has always been puzzling: Why should experienced "knowledge" exist at all if it cannot influence anything physical, and hence be of no use to the organisms that possess it. And how can something like an "idea", seemingly so different from physical matter, as matter is conceived of in classical mechanics, be created by, or simply *be*, the motion of physical matter?

The basic precepts of classical mechanics are now known to be fundamentally incorrect: they cannot be reconciled with a plenitude of empirical facts discovered and verified during the twentieth century. Thus there is no reason to demand, or believe, that those puzzling properties of the classically conceived world must carry over to the actual world, which conforms far better to the radically different precepts of quantum mechanics.

The founders of quantum theory conceived their theory to be a mathematical procedure for making practical predictions about future empirical/experiential findings on the basis of present empirical knowledge. According to this idea, quantum theory is basically about the evolution of knowledge. This profound shift is proclaimed by Heisenberg's assertion [27] that the quantum mathematics "represents no longer the behavior of the elementary particles but rather our knowledge of this behavior", and by Bohr's statement [28] that "Strictly speaking, the mathematical formalism of quantum mechanics merely offers rules of calculation for the deduction of expectations about observations obtained under conditions defined by classical physics concepts."

The essential need to bring "observations" into the theoretical structure arises from the fact that physical evolution via the Schrödinger equation, which is the quantum analog of the classical equations of motion, produces in general not a single evolving physical world that is compatible with human experience and observations, but rather a mathematical structure that corresponds to a smeared out mixture of increasingly many such worlds. Consequently, some additional process, beyond the one generated by the Schrödinger equation, is needed to specify the connection is between the physically described quantum state of the universe and experienced empirical reality.

This important connectivity is alien to the concepts of classical physics. Those concepts arose from—or were at least heavily reinforced by—the conceptual miniaturization of the celestial objects of astronomy and the solid terrestrial objects of normal observation. In those two regimes we, the observers, stand effectively apart from the system being observed and—under the conditions of the applicability of that classical physical theory—have no appreciable influence upon the behavior of the observed system. The classical concept of "the physical system" was thereby divorced from the concept of "being observed".

This classical separability the physical from the mental is not altered by miniaturization. However, there is no rational reason why this separability feature of the classical conceptualization of the physical world should continue to be useful or applicable when the brains of *we the observers* become included in what is being described physically. But how does scientific theory advance in a well-defined and useful way beyond the classical notion of mind-brain disjunction? How can science bring these two disparate kinds of descriptions together in a rationally coherent manner?

The founders of quantum mechanics achieved a profound advance in our understanding of nature when they recognized that the mathematically/physically described universe that appears in our best physical theory represents *not* the world of material substance contemplated in the classical physics of Isaac Newton and his direct successors, but rather a world of *"potentia"*, or *"weighted possibilities"*, for our future acquisitions of *knowledge* [29]. It is not surprising that an adequate scientific theory designed to allow us to predict correlations between our shared empirical findings should incorporate, as orthodox quantum mechanics does: (*1*), a natural place for *"our knowledge"*, which is both all that is really known to us, and also the empirical foundation upon which science is based; (*2*), an account of the process by means of which we acquire our *knowledge* of certain physically described aspects of nature; and (*3*), a statistical description, at the pragmatic level, of relationships between various features of the growing aspect of nature that constitutes "our knowledge".

What is perhaps surprising is the ready acceptance by most western-oriented scientists and philosophers of the notion that the element of chance that enters quite reasonably into the *pragmatic* formulation of physical theory, in a *practical* context where many pertinent things may be unknown to us, stems from an occurrence of raw pure chance at the underlying *ontological* level. Ascribing such capriciousness to the underlying basic reality itself would seem to contradict the rationalist ideals of Western Science. From a strictly rational point of view, it is, therefore, not unreasonable to examine the mathematical impact of tentatively accepting, *at the basic ontological level*, Einstein's dictum that: "God does not play dice with the universe", and thus to attribute the effective entry of pure chance at the practical level to our lack of knowledge of the *reasons* for the supposedly random choices that enter into the quantum dynamics to be what they turn out to be.

These supposedly random choices enter quantum mechanics only through certain "choices on the part of nature". These choices determine which of the potentialities generated by the mechanistic Schrödinger equation are actualized and experienced. The tentative assumption, here, is that the seeming randomness of these choices arises from the incompleteness of our knowledge of the conditions that determine what these choices will be, but that sufficient reasons for these choices do exist, and a proper task of science is to find out what some of these reasons are.

Implementing the Principle of Sufficient Reason

I make no judgment regarding the technical correctness of the purported evidence for the existence of the reported retrocausal phenomena. That I leave to the collective eventual wisdom of the scientific community. I am concerned here rather with essentially logical and mathematical issues, as they relate to the apparent view of some commentators that scholarly articles reporting the existence of retrocausal phenomena should be banned from the scientific literature, essentially for the reason articulated in the New York Times by Douglas Hofstadter, namely that the actual existence of such phenomena is irreconcilable with what we now (think we) know about the structure of the universe. But is it actually true that the existence of such phenomena would require a wholesale abandonment of basic ideas of contemporary physics.

That assessment is certainly not valid, as will be shown here. A limited, and intrinsically reasonable, modification of the existing orthodox quantum mechanics is sufficient to accommodate the reported data. Hence banning the publication of such works would block a possible important advancement in science that would constitute an empirically small but conceptually important correction to contemporary mainstream science. The issue in question is the validity of Einstein's opinion that the randomness invoked by orthodox quantum mechanics is not a fundamental feature of reality itself.

In order for science to be able to confront effectively purported phenomena that violate the prevailing basic theory, what is needed, or at least helpful, is an alternative theory that retains the empirically valid predictions of the currently prevailing theory, yet accommodates in a rationally coherent way the claimed new phenomena.

If the example of the transition from classical physics to quantum physics can serve as an illustration, in that case we had a beautiful theory that had worked well for 200 years, but that was incompatible with the new data made available by advances in technology. However, a new theory was devised that was closely connected to the old one, and that allowed us to recapture the old results in the appropriate special cases, where the effects of the nonzero value of Planck's constant could be ignored. The old formalism was by-and-large retained, but readjusted to accommodate the fact that properties that according to ordinary classical ideas were described by *numbers* that specified the actual numerical values of the properties, were represented at a more basic level by *actions,* which were related to the *measurement processes* by means of which the numerical values were empirically ascertained. Thus the active process by means of which *we find out about* certain pertinent numbers was brought explicitly into the dynamical theory. This restructuring that brings into the heart of the theory our actions of performing the measurements that produced the increments in our knowledge that constituted our empirical findings is closely tied to *a rejection of a basic classical presupposition*, namely the idea that basic physical theory should properly be primarily about connections between physically described material events, with experiential ramifications an inessential addendum. The founders of quantum theory insisted, in direct contrast, that their more basic physical theory was essentially pragmatic—i.e., was directed at predicting practically useful connections between empirical (i.e., experienced) events [30].

This original pragmatic Copenhagen QM was not suited to be an ontological theory, because of the movable boundary between the aspects of nature described in *classical* physical terms and those described in *quantum* physical terms. It is certainly not ontologically realistic to believe that the pointers on observed measuring devices are built out of *classically* conceivable electrons and atoms, etc. The measuring devices, and also the bodies and brains of human observers, must be understood to be built out of quantum mechanically described elements. This is what allows us to understand and describe many observed properties of these physically described systems, such as their rigidity and electrical conductance. The aspects of quantum mechanics that describe our observations is more accurately called a description of the experiential aspects, which can make use of classical concepts as aids to our descriptions of our experiences.

Von Neumann's analysis of the measurement problem allowed the quantum state of the universe to describe the entire physically described universe: everything that we naturally conceive to be built out of atomic constituents and the fields that they generate. This quantum state is described by assigning mathematical properties to space-time points (or tiny regions). There is a deterministic law, the Schrödinger equation, that specifies the mindless, essentially mechanical, evolution of this quantum state. But this quantum mechanical law of motion generates *a huge continuous smear of worlds of the kind that we actually experience.* For example, as Einstein emphasized, the position of the pointer on a device that is supposed to tell us the *time* of the detection of a particle produced by the decay of a radioactive nucleus, evolves, under the control of the Schrödinger equation, into a *continuous smear of positions corresponding to all the different possible times of detection*; not to a single position, which is what we observe [31]. And the unrestricted validity of the Schrödinger equation would lead, as also emphasized by Einstein, to the conclusion that the moon, as it is represented in the theory, would be smeared out over the entire night sky, until the first observer of it, say a mouse, looks.

How do we understand this huge disparity between the representation of the universe evolving in accordance with the Schrödinger equation and the empirical reality that we experience?

An adequate physical theory must include a logically coherent explanation of how the mathematical/physical description is connected to the experienced empirical realities. This demands, in the final analysis, a theory of the mind-brain connection: a theory of how our idea-like knowings are connected to our evolving physically described brains.

The micro-macro separation that enters into Copenhagen QM is actually a separation between what is described in quantum mechanical physical terms and what is described in terms of *our experiences*—expressed in terms of our everyday concepts of the physical world, refined by the concepts of classical physics [28, Sect. 3.5].

To pass from *quantum pragmatism* to *quantum ontology* one can treat all *physically described* aspects quantum mechanically, as Von Neumann did. He effectively transformed the Copenhagen pragmatic version of QM into a potentially ontological version by shifting the brains and bodies of the observers—and all other physically described aspects of the theory—into the part described in quantum mechanical language. The entire physically described universe is treated quantum mechanically, and *both our knowledge*, and *the process by means of which we acquire our knowledge about the physically described world,* are elevated to essential features of the theory, not merely postponed, or ignored! Thus certain aspects of reality that had been treated superficially in the earlier classical theories—namely "our knowledge" and "the process by means of which we acquire our knowledge"—were now incorporated into the theory in a detailed way.

Specifically, each acquisition of knowledge was postulated to involve, first, a "choice of probing action executed by an observing agent", followed by "a choice on the part of nature" of a response to the agent's request (demand) for this particular piece of experientially specified information.

This response on the part of nature is asserted by orthodox quantum mechanics to be controlled by *random chance*, by *a throw of nature's dice*, with the associated probabilities specified purely in terms of physically described properties. These "random" responses create a sequence of collapses of the quantum state of the universe, with the universe created at each stage concordant with the new state of "our knowledge".

If Nature's choices conform strictly to these orthodox statistical rules then the results reported by Bem cannot be accommodated. However, if nature is not capricious—if God does *not* play dice with the universe—but Nature's choices have sufficient reasons, then, given the central role of "our knowledge" in quantum mechanics, it becomes reasonable to consider the possibility that Nature's choices are not completely determined in the purely mechanical way specified by the orthodox rules, but can be *biased* away from the orthodox rules in ways that depend upon the character of the knowledge/experiences that these choices are creating. The results reported by Bem can then be explained in simple way that elevates the individual "choices on the part of nature" from "choices that are determined by absolutely nothing at all", to "choices that arise from relevant conditions that include the experienced emotions of biological agents."

In classical statistical physics such a biasing of the *statistics* would not produce the appearance of *retrocausation*. But in quantum mechanics it does! The way that the biasing of the forward-in-time quantum causal structure leads to *seemingly* "retrocausal" effects will now be explained.

Backward in Time Effects in Quantum Mechanics

The idea that choices made now can influence what has already happened needs to be clarified, for this idea is, in some basic sense, incompatible with our idea of the meaning of time. Yet the empirical results of Wheeler's delayed-choice experiments [32], and the more elaborate delayed-choice experiments of Scully and colleagues [33] are saying that, *in some sense*, what we choose to investigate now can influence what happened in the past. This backward-in-time aspect of QM is neatly captured by an assertion made in the recent book "The Grand Design" by Hawking and Mlodinow: "We create history by our observations, history does not create us" [34].

How can one make rationally coherent sense out of this strange feature of QM?

I believe that the most satisfactory way is to introduce the concept of "process time". This is a "time" that is different from the "Einstein time" of classical deterministic physics. That classical time is the time that is joined to physically described space to give classical Einstein space-time. (For more details, see my chapter in "Physics and the Ultimate Significance of Time" SUNY, 1986, Ed. David Ray Griffin. In this book three physicists, D. Bohm, I. Prigogine, and I, set forth some basic ideas pertaining to time [35].)

Orthodox quantum mechanics features the phenomena of collapses (or reductions) of the evolving quantum mechanical state. In orthodox Tomonaga-Schwinger relativistic quantum field theory [36, 37, 38], the quantum state collapses not on an advancing sequence of constant time surfaces (lying at a sequence of times $t(n)$, with $t(n + 1) > t(n)$, as in non-relativistic QM), but rather on an advancing sequence of *space-like surfaces* $\Sigma(n)$. (For each n, every point on the spacelike surface $\Sigma(n)$ is spacelike displaced from every other point on $\Sigma(n)$, and every point on $\Sigma(n + 1)$ either coincides with a point on $\Sigma(n)$, or lies in the open future light-cone of some points on $\Sigma(n)$, but not in the open backward light-cone of any point of $\Sigma(n)$.)

At each surface $\Sigma(n)$ a projection operator $P(n)$, or its complement $P'(n) = I - P(n)$, acts to reduce the quantum state to some part of its former self!

For each surface $\Sigma(n)$ there is an associated "block universe", which is defined by extending the quantum state on $\Sigma(n)$ both forward and backward in time via the unitary time evolution operator generated by the Schrödinger equation. Let the index n that labels the surfaces $\Sigma(n)$ be called "process time". Then for each instant n of process time a "new history" is defined by the backward-in-time evolution from the newly created state on $\Sigma(n)$.

This new "effective past" is the past that smoothly evolves *into the future the quantum state (of the universe) that incorporates the effects of the psycho-physical event that just occurred.* As far as current predictions about the future are concerned it is *as if* the past were the "effective past": the former *actual* past is no longer pertinent because it fails to incorporate the effects of the psycho-physical event that just occurred.

In orthodox QM each instant of process time corresponds to an "observation": the collapse at process time n reduces the former quantum state to the part of itself that is compatible with the increased knowledge generated by the new observation. This sequential creation of a sequence of new "effective pasts" is perhaps the strangest feature of orthodox quantum mechanics, and the origin of its other strange features.

The *actual* evolving physical universe is generated by the always-forward-moving creative process. It is forward-moving in the sense that the sequence of surfaces $\Sigma(n)$ advances into the future, and at each instant n of process time some definite, never-to-be-changed, psycho-physical events happens. But this forward-moving creative process generates in its wake an associated sequence of effective pasts, one for each process time n. *The conditions that define the effective past associated with process time n change the preceding effective past imposing a "final" condition that represents what happened at process time n.* It is this "effective past" that evolves directly into the future, and is the past that, from a future perspective, has smoothly evolved into what exists "now". The actual past is not relevant to a history of the universe that starts from now and looks back, and projects smoothly into the immediate future.

The "histories" approach to quantum physics focuses attention on histories, rather than the generation of the profusion of incompatible possibilities. Both the effective past and the history associated with process time n depend upon which experiment is performed at time n, and in quantum mechanics that choice of which experiment is performed at process time n is not determined by the quantum state at process time n: it depends upon the agent's "free choice" of which probing action to initiate, where the word "free" specifies precisely the fact that this choice on the part of the agent is not determined by the known laws of nature.

Two key features of von Neumann's rules are mathematical formalizations of two basic features of the earlier pragmatic Copenhagen interpretation of Bohr, Heisenberg, Pauli, and Dirac. Associated with each observation there is an initial "choice on the part of the observer" of what aspect of nature will be probed. This choice is linked to an empirically recognizable possible outcome "Yes", and an associated projection operator $P(n)$ that, if it acts on the prior quantum state ρ, reduces that prior state to the part of itself compatible with the knowledge gleaned from the experiencing of the specified outcome "Yes".

The process that generates the observer's choice of the probing action is not specified by contemporary quantum mechanics: this choice is, *in this very specific sense*, a "free choice on the part of the experimenter." Once this choice of probing action is made and executed, then, in Dirac's words, there is "a choice on the part of *nature*": nature randomly selects the outcome, "Yes" or "No" in accordance with the statistical rule specified by quantum theory. If Nature's choice is "Yes" then $P(n)$ acts on the prior quantum state ρ, and if nature's answer is "No" then the complementary projection operator $P'(n) = I - P(n)$ acts on the prior state. Multiple-choice observations are accommodated by decomposing the possibility "No" into sub-possibilities "Yes" and "No".

Mathematical Details

The description of orthodox quantum mechanics given above is a didactic equation-free account of what follows from the equations of quantum measurement theory. Some basic mathematical details are given in this section.

The mathematical representation of the dynamical process of measurement is expressed by the two basic formulas of quantum measurement theory:

$$\rho(n+1)_Y = \frac{P(n+1)\rho(n)P(n+1)}{Tr(P(n+1)\rho(n)P(n+1))},$$

and

$$<P(n+1)>_Y = Tr(P(n+1)\rho(n)P(n+1)) = Tr(P(n+1)\rho(n)).$$

Here the integer *"n"* identifies an element in the global sequence of probing "measurement" actions. The symbol $\rho(n)$ represents the quantum state (density matrix) of the observed physical system (ultimately the entire physically described universe, here assumed closed) immediately *after* the nth measurement action; $P(n)$ is the (projection) operator associated with answer *"Yes"* to the question posed by the nth measurement action, and $P'(n) = I - P(n)$ is analogous projection operator associated in the same way with the answer *"No"* to that question, with *"I"* the unit matrix. The formulas have been reduced to their essences by ignoring the unitary evolution *between* measurements, which is governed by the Schrödinger equation.

The expectation value $<P(n + 1)>_Y$ is the normal orthodox probability that nature's response to the question associated with $P(n + 1)$ will be *"Yes"*, and hence that $\rho(n + 1)$ will be $\rho(n + 1)_Y$. In the second equation I have used the defining property of projection operators, $PP = P$, and the general property of the trace operator: for any X and Y, $Tr(XY) = Tr(YX)$. (The trace operation Tr is defined by: $Tr(M) =$ Sum of the diagonal elements of the matrix M).

Of course, one cannot know the density matrix ρ of the entire universe. The orthodox rules tell us to construct a "reduced" density matrix by taking a partial trace over the degrees of freedom about which we are ignorant, and renormalizing. This eliminates from the formulas the degrees of freedom about which we are ignorant.

The trace operation is the quantum counterpart of the classical integration over all of phase space. The classical operation is a summation that gives equal a priori weighting to equal volumes of phase space. That is the weighting that is invariant under canonical transformations, which express physical symmetries. The quantum counterparts of the canonical transformations are the unitary transformations, which leave the trace unchanged. Thus the orthodox trace rules are the rational way to give appropriate weights to properties about which we have no knowledge, namely by assuming that properties related by physical symmetries should be assigned equal a priori weights.

All this is just orthodox quantum mechanics, elaborated to give a rationally coherent ontological account compatible with the standard computational rules and predictions [39].

But the assumption that nature gives equal weights to properties that we, in our current state of scientific development, assume should be given equal weights, does not mean that nature itself must give such properties equal weight. Two states of the brain that are assigned equal statistical weight by the orthodox trace rule may be very different in the sense that one corresponds to a meaningful, coherent, pleasing experience and the other does not. Classical mechanics postulates that experiential qualities, per se, can make no difference in the flow of physical events. But, since quantum mechanics places experiences in a much more central role than classical mechanics, there is no rationally compelling reason to postulate in quantum mechanics that nature, in the process of choosing outcomes of empirical questions posed by agents, must be oblivious to the experiential aspects of reality. That issue should be settled by empirical findings, not by classical-physics-based prejudice.

Consider a situation in which: (1), an agent (the participant) observes a property that corresponds to a projection operator P; and (2), a dynamically independent random number generator (RNG) creates either the property represented by the projection operator Q, or the property represented by the complementary property $Q' = (I - Q)$. Suppose at some time after these properties have been created they are still confined to two different systems that have never interacted, so that $PQ = QP$, and $\rho = \rho(P)\,\rho(Q)$. Then the probability of getting the answer (PYes), given that (QYes) occurs, is:

$$Trace\,PQ\rho / Trace\,Q\rho = Tr\,P\rho(P) / Tr\,\rho(P),$$

which is independent of Q: the probability of P does not depend on what the dynamically independent RNG does.

Suppose, now, the two systems interact later, beginning at time t, then the propagation to a final later time t', at which time an observable corresponding to projection operator R is measured. The predicted statistical correlation between the outcomes of the measurements associated with P and the outcomes of measurements associated with Q will now normally depend upon whether the outcome of the final measurement is the "Yes" associated with projection operator R, or the "No" associated with the projection operator $(I - R)$. But the orthodox rules ensure that if one sums the contributions from R and $(I - R)$, using the weights prescribed by those orthodox rules, then this dependence on R will drop out. If, on the other hand, the probabilities of nature's choices between R and $(I - R)$ differ from the orthodox ones, then, after Nature's biased choice, the theory predicts observable correlations between the outcomes of the measurements of P and Q: the outcomes of these measurements that are predicted to be *un*correlated by orthodox quantum mechanics will now be predicted to be correlated. This change in the predictions arise from the contributions of some extra weighted histories brought in by Nature's biased choice, and the absence of some other weighted histories.

Applications to Bem's Experiments

All nine of Bem's experiments have the following general form: First, in each instance in a series of experimental instances, the participant is presented with some (in most cases emotionally neutral) options, and picks a subset of these options as 'preferred'. These preferences are duly recorded. *Later*, for each instance, an emotional stimulus is applied to the participant. The stimulus, and the way it is applied to the participant, is determined by some random number generators (RNGs). These RNGs are, according to both classical and quantum ideas, dynamically independent of the participant's earlier actions. But Bem's empirical result is that the probability that an option is preferred by the participant at the earlier time depends upon choices made later by the RNGs.

This finding seems to suggest that either the believed-to-be dynamically independent RNGs are being influenced in a mysterious and complex way by the participant's earlier actions; or the participant's earlier actions are being affected in a complex retrocausal (backward-in-time causal) way by the choices made by RNGs.

The kinds of actions made by the participant, and by the RNGs, vary greatly over the nine experiments. But, from a quantum standpoint, one single presumption explains all of the reported results, and explains them all in a basically forward-in-time causal way, without any mysterious influence of the participant's choice of preference on the RNGs. This presumption is that the choices on the part of nature, which are essential elements of orthodox quantum mechanics, are slightly biased, relative to the orthodox quantum statistical rules, in favor of the actualization of positive feelings in the mind of the participant, or, in other cases, against the actualization of negative feelings.

For example, in the first Bem experiment the participant is shown two similar screens, L and R, and is told that behind one screen lies a picture, and behind the other lies the image of a blank wall. S/he is instructed to choose a "preferred" screen, P (either L or R) behind which s/he *feels* the picture lies. *After* the participant's preference P, either L or R, is recorded, a first random number generator, RNG1, chooses a "target" screen T (either L or R), and assigns a *picture* to target screen T, and an image of a blank wall to the other screen. A second random number generator, RNG2, decides, with equal probabilities, whether the picture will be "Erotic" or "Neutral" (The stimulus type S is either E or N). What has been determined by the RNGs to lie behind the preferred screen P is then shown to the participant.

Bem's empirical result is that the participants choose, more often than orthodox quantum mechanics (or classical statistical mechanics) predicts, the screen behind which *will lie* an erotic picture, but prefers L and R with equal probability if RNG2 chooses a "neutral" picture.

If the well-tested random number generators are working as they normally do then this empirical result would appear to be a case of retrocausation (causal action backward in time): the choices made *later* by the two RNGs are influencing the subject's *earlier* choice between L and R. The idea that the present can *actually* change the past would introduce huge conceptual problems into quantum mechanics, and would require a major re-thinking and re-construction of the entire theory, centering on the problem of how to retain the massive body of valid predictions. It would bring into play Hofstadter's observation that the whole edifice of contemporary theory would be toppled. Changing the past would often cause big changes in the present. How could one salvage the predictions of the tremendously successful orthodox physical theory?

An alternative possibility is that RNG2, which chooses between "erotic" and "neutral", is being influenced by the participant's earlier choice between L and R, so that the screen behind which the participant looks will tend to be erotic. But this should occur only if RNG1 chooses "picture" not "blank wall". Moreover, the key variable is an emotional response on the part of the subject that has not yet occurred when the supposed action of the subject's earlier choice between two *neutral images* upon RNG2s choice is supposed to occur. That emotional response is fixed by an arbitrary mechanism, designed by the experimenters, that has not yet been brought into play.

These problems constitute major difficulties. But Bem's results are explained in natural, rational, essentially forward-causal way, without any apparent difficulties, provided *Nature's choice of the participant's final experience*—a choice that is an absolutely essential element of orthodox quantum theory—favors, relative to the statistical predictions of orthodox quantum mechanics, the occurrence of positive (pleasing) experiences and disfavors the occurrence of negative (displeasing) feelings. If such a biasing of Nature's choices were to occur, then the observed greater likelihood of the participant's choosing the screen, L or R, behind which an erotic picture *will lie* would arise directly from the enhanced likelihood that nature will actualize an erotic experience rather than an experience of a neutral picture or a blank wall.

In this experimental set up an erotic experience can occur only if $P = T$ and $S = E$: the participant's earlier choice of the between L and R must agree with the later choice of RNG1 between L and R, since otherwise the participant will see only a blank wall, and even if $P = T$, the choice of stimulus S must be E, since otherwise the participant will see a neutral picture.

A compact way of stating this explanation is to say that the quantum histories [defined by the sequences of choices (P, T, S, F) leading to the final experience $F = +$, or $F = -$] that lead to $F = +$ are more likely to occur than the rules of orthodox quantum mechanics predict. Only those histories in which the two L/R choices agree ($P = T$) can lead to an erotic experience, because if these two choices disagree the participant will see a blank wall. But this enhancement will occur only in the subset of histories in which $S = E$.

In Bem's Experiment 2, "Precognitive Avoidance of (Subliminal) Negative Stimuli", a sequence of similar pairs of neutral pictures is shown to the participant, who chooses a 'preferred' picture from each neutral pair. After each such recorded choice of preference P, a RNG1 makes a random choice of one picture from the initial pair. The picture chosen by RNG1 is called the 'target' T. Then the apparatus flashes a *subliminal* picture, the stimulus S, that is positive, $S = +$, if $T = P$, but is highly negative, $S = -$, if the preferred neutral picture P is *not* the subsequently randomly chosen target picture T.

The normal idea of forward causation does not allow this random choice of target, and the associated application of a stimulus, both of which occur *after* the recorded choice of preference, to affect, in any instance, the participant's previously recorded choice of preference between two matched neutral pictures. Yet Bem's predicted and empirically validated result is that the picture P preferred at an earlier time by a participant is more likely to be the subsequently chosen target picture T than the subsequently chosen non-target, even though the choice between target and non-target was 50–50 random, and was made only later. The non-targeted pictures, which are, according to Bem's empirical findings, less likely to be preferred than chance predicts, are the pictures that occur in conjunction with the later subliminal application to the participant of highly unpleasant pictures. Hence they should lead to unpleasant participant feelings and should therefore, according to the present hypothesis, be less likely than chance predicts to be selected by Nature's choice to become an actually experienced outcome:

$$< (P, T \text{ not } P, S-, F-) > \; < \; <P, T = P. S+, F+ > .$$

This experimental protocol is quite different from the protocol of the first experiment. In the first experiment the stimulus that was applied later to the participant was independent of the participant's earlier choice of preference, whereas in experiment 2 the stimulus that is applied later to the participant depends upon the earlier choice of preference. Moreover, the stimulus was supraliminal in the first experiment but subliminal in the second experiment.

Nevertheless, the apparently retrocausal effect in the second experiment follows from the same quantum assumption as before, namely that Nature's choice of which final experience actually occurs has a tendency to increase the likelihood of positive, and diminish the likelihood of negative, final *feelings* of the participant. In experiment 2 the effect of this biasing is to diminish the likelihood of instances in which the *final feeling* of the participant is negative, due to the earlier application to the participant of an (albeit subliminal) highly negative stimulus.

Bem's experiments 3 and 4 are "Retrocausal Primings". Unlike the first two experiments, they do not involve matched neutral pairs between which the participant must choose. Rather, each instance now involves a single picture, which is emotionally either positive or negative. This non-neutral picture is shown to the participant, who responds by pressing a first or second button according to whether s/he feels the picture to be pleasing or not. The *time* that it takes for the participant to react to the picture is recorded. *Then,* a 'word' is selected by a RNG, and is (supraliminally) shown to the participant. The previously recorded reaction time turns out to be shorter or longer according to whether feeling of the word is "congruent" or 'incongruent" to the feeling of the picture experienced earlier. For example, the word "beautiful" is congruent to the picture of Grace Kelly, but incongruent to a picture of Bela Lugosi as Count Dracula.

There is also a 'normal' version of the experiment in which the word chosen by the RNG is displayed *before* the participant chooses his preference. Bem's experimental set-up is one for which, also in the 'normal' version, the recorded reaction time is shorter or longer according to whether feeling of the word is "congruent" or "incongruent" to the feeling of the picture shown earlier.

The question at issue is: How, in the retrocausal version, can the *reaction time*, which was recorded *earlier,* depend on which word was randomly selected *later*?

This empirical finding is explained by an assumed biasing of "Nature's choice" of the participant's final feeling that favors congruency in the flow of experience over incongruency. Such a putative biasing of Nature's choice has the effect of adding to the effective past, after nature's biased choice, some extra histories that lead to the mentioned positive feelings, or to the subtraction of histories that lead to analogous negative feelings. These differences in the set of contributing histories, in accordance with the nature of the feeling induced by the stimulus word, have an effect on the quantum state of the participant's brain *during the process of his or her choosing between positive and negative pictures.* This effect on the brain during that period is similar to the effect of applying a similar stimulus before the participant's choice of response. In both cases the "effective past" state of the brain of the participant during his or her process of choosing a response is changed in essentially the same way: it is not important whether the change in the effective state of the participant's brain, during the process of choosing his or her preference, comes from changes in the earlier or later boundary condition on that "effective past" state of the brain. The key point is that, as discussed in earlier sections, the "effective past" incorporates the conditions imposed by the occurrence of the *final* outcome! A "history" starts from what is now known, and extends backward from the known present, which depends on nature's most recent choice.

The next three experiments relate to the well-known phenomena of "habituation". The participant is again shown an emotionally matched pair of pictures, and is asked which one s/he prefers. The two matched pictures are both strongly negative, both strongly positive (erotic), or both essentially neutral in the first, second, and third experiments, respectively. (I have slightly reorganized Bem's data in this way for logical clarity, and ignored some inconclusive data with small statistics in which certain later stimuli were supraliminal.) After the participant makes a binary recorded choice of preference, an RNG chooses one of the two similar pictures as target, and the targeted picture is subliminally flashed several times. The subliminal re-exposures, made after the participant's choice of preference of the targeted emotion-generating picture, have the effect of reducing, in the case of the positive pairs of pictures, and increasing in the case of negative pairs of pictures, the fraction of instances in which the (previously) preferred picture was the target rather than the non-target: the effective positivity/negativity of the targeted (and hence repeatedly subliminally represented) pictures was reduced. This is explained by a *reduction* in the emotional intensity of the participant's final feeling, caused by the repeated re-exposure to the highly emotional pictures, and the attendant *diminuation* of the biasing of Nature's choices.

In the final two (memory) experiments the participant is exposed to a sequence of 48 common everyday nouns, and is then tested see which words s/he remembers. Afterwards, 24 of the original set of words are randomly chosen to be 'targets, and then, in a sequence of computer-controlled actions, the participant is repeatedly re-exposed to each of the target words, but to none of the non-target words. It is subsequently found that among the recalled words there are more target words than non-target words. This is explained if Nature's choice of the participant's final feelings favors the feel of congruent streams of conscious experiences over the feel of less congruent ones.

Of course, the actual past has not been changed. If the participant had been graded "pass or fail" according to the number of words recalled, then his grade would not depend upon what happened later. In that earlier test each initially presented word would either be recalled or not recalled, and s/he would pass of fail on the basis of the number recalled. Suppose 60% are recalled and 40% are not recalled, which is failing. Given any thus-determined particular individual outcome one could find out, after the experiment, whether it was targeted or not. Suppose, as an extreme case, that targeting is extremely effective and that every targeted word is recalled, and no non-targeted word is recalled. That would give a perfect positive correlation between recall and targeting, but would not change the grade from failing to passing. If Nature's choices can be biased relative to the orthodox pre-dictions in the way indicated by the Bem experiments, then empirically observed correlations between recorded past events can be a consequence of the actualizing capacity of Nature's biased choices, rather than an expression of correlations that existed prior to Nature's choices of which of the quantum-generated potentialities to make actual. This would render the past unchangeable, but the future somewhat dependent upon our desires, and the congruency of thoughts.

All of Bem's reported results are thus explained by a single presumption, namely that Nature's choices, rather than being strictly random, in accordance with the rules of contemporary orthodox quantum mechanics, are sometimes slightly biased, relative to the predictions of orthodox quantum mechanics, in favor of outcomes that feel pleasing, and against outcomes that feel displeasing.

This explanation is "scientific", in the sense that *it can be falsified*. If the output of the RNGs were to be observed by an independent observer, *before* the RNG-chosen action is made on the participant, then the biasings reported by Bem should disappear, because Nature's choice would then be about the possible experiences of the independent observer rather than about those of the participant.

A more elaborate test would be to have two participants doing the experiment on the same sequence of pictures, with reversed polarities. A dependence upon who first experiences the output of the RNG would, if it were to occur, constitute spectacular support for the notion that our experiences really do influence the course of physically described events, rather than being merely causally inert by-products of a process completely determined by purely physical considerations alone.

In the above discussion I have treated all of the RNGs as true quantum-process-based random number generators. In some of the experiments the RNG was actually a pseudo-random number generator, a PRNG. In principle a PRNG is, in these experiments, just as good as a true RNG, unless *at the time of its effective action* some real observer *actually knows* everything needed to specify what the pseudo-random choice must be. Unless the outcome is actually specified by what is actually currently known by observing agents, the outcome is, within this orthodox framework, effectively undetermined.

Conclusion

Bem's seemingly backward-in-time causal effects can be explained within a quasi-orthodox forward-in-time quantum mechanics. In this variation of orthodox theory, Nature's "random" choices of which outcomes of measurements to actualize are slightly biased away from the random choices prescribed by the orthodox theory in favor of outcomes that actualize positive feelings of the participants, and against outcomes that actualize negative feelings of the participants.

Acknowledgements This work was supported by the Director, Office of Science, Office of High Energy and Nuclear Physics, of the U.S. Department of Energy under contract DE-AC02-05CH11231

Appendix E
The Quantum Conception of Man[1]

Quantum Mechanics, Consciousness, Spooky Action-at-a-distance, Bell's Theorem, and Free Will

Each of these topics is a deep subject about which much has been written. I intend to describe here tonight my own view of how these various elements fit together to form a rationally coherent understanding of the world that we human beings inhabit, and of our role within it.

I have been thinking about the matters for more than 50 years. Already in 1958 I was working on them in Zurich with Wolfgang Pauli, a principal founder of quantum mechanics.

When he unexpectedly died, I read von Neumann's book on these matters, and then wrote an essay to myself entitled:

Mind, Matter, and Quantum Mechanics,

which eventually developed into a 1993 book of the same title.

In the seventies I worked on these matters in Munich with Werner Heisenberg and wrote in the American Journal of Physics a seminal article entitled "The Copenhagen Interpretation of Quantum Mechanics". I have continued to think and write about these matters.

The first topic is quantum mechanics. In order to understand Quantum Mechanics, it is important contrast it with what came before it, namely "Classical Mechanics". Classical Mechanics was created by Isaac Newton, who said "It seems to me probable that God in the beginning created matter in solid, massy, hard, impenetrable movable particles." These particles can interact locally by contact, like billiard balls. But they can also act upon each other by gravitational attraction. In Newton's theory gravity acts instantaneously over astronomical distances. Thus already at the beginning of modern science we encounter a "Spooky action-at-distance."

However, about 200 years later, Maxwell created a wave theory of the interactions between charged particles: The information carried by such a wave could be

---

[1]Talk presented to the Mount Diablo Astronomical Society, January 27, 2015.

transmitted no faster than a certain maximal speed that could be calculated, and turned out to be the empirically measured speed of light. Maxwell's waves were Light Waves.

A few years later Einstein, in his theory of relativity, re-formulated all of classical physics so that no physical structure could transmit information faster than the speed of light. Thus Einstein banished Spooky action-at-a-distance from classical physics.

A second main property of classical physics is physical determinism, which says that all physically described properties are completely determined by prior physically described properties.

This property is also called the "Causal Closure of the Physical". It means that the behavior of your physical body was completely predetermined already at the birth of the universe:

This property turns you into a mechanical automaton, and converts your intuition that your conscious "free will" can influence your bodily behavior into a pervasive illusion.

However, that conclusion does not carry over to the quantum world.

Quantum Mechanics.

The quantum story begins with Max Planck's discovery at the beginning of the twentieth century that Light Waves have a corpuscular character: The transfer of energy between light waves and physical particles seems to occur in finite "chunks", called "quanta". The sizes of these "chunks" are directly proportional to the frequency of the light.

Atomic physicists then tried to construct a conception of atoms that would account for all the existing empirical facts. They tried at first to use the same kinds of ideas that Newton had used to explain the motions of the planets circling about the sun to explain, now, the motions of the electrons circling about the atomic nucleus. A 25 year struggle showed that that idea would not work. Then Heisenberg, and also Schrödinger, working independently, discovered the equation that made it all work. And that equation, properly generalized, covered not only single atoms, but also collections of arbitrarily large numbers of atoms, and hence large hard objects such as tables and chairs, and also, among other things, the measuring devices that are used to measure atomic properties. The theory gives predictions about, for example, the location on a dial of visible "Pointer". The position of this pointer reports to us human observers the value of some microscopic property of the system being examined.

The problem, however, is that this straight-forward prediction does not agree with human perceptions. The predicted position of the pointer turns out to be a smear over a large range of possible values, whereas the human observers see the position of the pointer confined, within small errors, to some tiny region of the dial.

Thus the basic problem is:

How are we to deal with this sharp disagreement between the quantum laws, which in principle ought to control the evolving state of the (interacting) atomic constituents of the world, with our perceptions of the world composed of those constituents?

Niels

The solution offered by quantum theory is expressed in Bohr's oft-repeated dictum:

In the drama of existence we are ourselves both actor's and spectators,

and in John Wheeler's likening of the quantum process of measurement to the game of twenty questions.

The details of this solution are most clearly spelled out in John von Neumann's rigorous reformulation of Copenhagen Quantum Mechanics. He explicitly introduces into the quantum dynamics, in addition to the normal quantum dynamical process, which he calls Process II, another dynamical process that he calls Process I.

This Process I converts the "Mental Observer" from a causally inert Spectator to a causally efficacious Actor. This Process I action has two phases. In the first phase the observer's mental aspect, his "ego" in von Neuman's terminology, poses a question: "Will my perception be P, where P is a classically described perception." In the second phase "Nature chooses and implements a psycho-physically described response, "Yes" or "No" to the observer's query.

The two main points are, first, that the observer's mental aspects are given a certain physically effective dynamical role in the evolution of the physically described universe—and, second, that a globally effective "Nature" produces an instantaneous global collapses that reinstate "Spooky action-at-a-distance".

This active dynamical role of the "Ego", even though it is only to instigate probing physical actions, is sufficient to allow, by means of rigorously specified basic quantum mechanical properties alone, a person's mental intentions to influence that person's bodily behavior in the mentally intended way. Quantum mechanics thus explains how your free-willed mental choices can be causally effective in the physical world!

It must be mentioned that in the late 1940s physicists (Tomonaga/Schwinger) created "Relativistic Quantum Field Theory", which allows all of the *empirical* consequences of Einstein's theory of relativity to be maintained in spite of the underlying spooky actions-at-a-distance associated with measurements.

Personal and Social Benefits of the Rescue of Free Will.

1. According to classical mechanics, your mental willful efforts can make no difference in the physically described world. If you are a rational person who bases your beliefs about the world upon science, then a belief in classical mechanics is debilitating, for it rationally causes you to believe that any effort you might make to improve your life or the lives of others is completely futile. On the other hand, your updated knowledge of the quantum mechanical character of the world is empowering because it lends scientific support to your essential-to-life, and experience-based, intuition that actions initiated by your value-based efforts can tend to bring pass that which you personally value.

2. Our legal system is based on the idea of personal responsibility for one's physical actions. But, according to classical mechanics, every physical action was predetermined at the birth of the universe. A person cannot rationally be held responsible for physical actions that were physically pre-ordained at the

birth of the universe. Quantum mechanics, on the other hand, does not entail any such physical predetermination, and thereby evades the classical-mechanics-based challenge to the rationality of our justice system!

Bell's Theorems.

Einstein banished Spooky action-at-a-distance from classical mechanics. His reaction to the instantaneous action at a distance that occurs in standard quantum mechanics was to agree with the founders of quantum mechanics that the rules of quantum mechanics should be viewed as mere practical computational procedures that allow scientists to make reliable predictions about future human experiences on the basis of their past experiences/perceptions. But Einstein believed that behind these merely statistical rules should lie a "reality" that likewise involves no Spooky action-at-a-distance.

Already in classical mechanics one can draw a distinction between a statistical state of a system and the underlying "real" possible states of that system: The statistical state is represented as a sum of terms each of which is a product of a positive weight factor times a possible "real" state.

John Stuart Bell formulated Einstein's position as the assertion that each of the statistically interpreted states of quantum mechanics can be expressed as a sum of terms each of which is a product of a positive weight factor times a possible "real" state that, in accordance with Einstein's intuition, allows no faster-than-light-action-at-a-distance. Bell and his associates proved many theorems that showed that no such decomposition is possible.

Those theorem's address one possible formulation of Einstein's position, but not the general question of whether the various empirical predictions of quantum mechanics can be satisfied if all spooky actions-at-distance are banned, in the sense that (in the standard example of a pertinent experiment, proposed by David Bohm) for each of the two alternative possible choices of which property is measured in a region the outcome there is independent of which experiment is freely chosen and performed at essentially the same time very far away. That is a cleaner formulation of Einstein's stated position, and it can be shown that such a banishing of Spooky actions cannot be reconciled with four basic and empirically well-validated predictions of quantum mechanics.

This result shows that Spooky actions cannot be banned, and hence that a materialistic conception of the physically described aspects of the world is incompatible with the empirical facts!

Appendix F
Mind, Brain, and Neuroscience

Introduction

The currently dominant theories of the connection of our conscious thoughts to our physical brains are based on the principles of classical mechanics. But those theories have achieved essentially no success in answering the "hard" question of how things as conceptually disparate as our conscious thoughts and classically conceived matter can combine together to form psychophysical human beings.

Yet classical mechanics is known to be empirically false. It has been replaced at the fundamental level by quantum mechanics. The primary difference between these two theories is that the classical mechanics never mentions our experiences whereas quantum mechanics is fundamentally about them, as Niels Bohr has often emphasized in statements such as:

> In our description of nature the purpose is not to disclose the real essence of phenomena, but only to track down as far as possible the multifold aspects of our experience [I, 18].

Thus our conscious experiences are the fundamental realities in quantum mechanics, whereas classical mechanics leaves them completely out.

It is therefore manifestly obvious that if a rational understanding of the mind-brain connection is being sought then quantum mechanics is the better theory to use. But why, then, are neuroscientist not using it?

The answer, I believe, it is simply that neuroscientists have not been shown how to do so. They have not been shown how to use the quantum mechanical model of the human person to compute, for example, the measured in vivo brain response to an associated mental choice.

My purpose in this talk is to illustrate how this is done, and compare the results to recent pertinent neuroscience data.

This example illustrates the quantitative workings of the quantum mechanical explanation of the influence of conscious intentions on in vivo brain activity.

© Springer International Publishing AG 2017
H.P. Stapp, *Quantum Theory and Free Will*,
DOI 10.1007/978-3-319-58301-3

Classical Description

An important feature of the seismic shift from classical to quantum mechanics is that the descriptive concepts of the earlier classical mechanics do not drop out, or fade away, but are transferred from the material reality that was supposed to lie *behind* our experiences to our experiences themselves. Thus in the words of Niels Bohr:

> ...it is important to recognize that in every account of physical experience one must describe both the experimental conditions and the observations by the same means of communication that is used in classical physics (II, p. 88).

Von Neumann's Solution to the Quantum Measurement Problem

The immediate consequence of this transfer of classical description to the mental realm is that, in practical measurement situations, the scientist is instructed to divide the world by a cut, called the Heisenberg cut, such that big things directly observed by observers are placed above the cut, and are described in terms of the concepts of classical mechanics, while things lying below the cut are described in quantum mechanical terms.

This rule was imprecise and ambiguous, and led to the so-called "measurement problem." John von Neumann, on the basis of a detailed mathematical examination, resolved this problem by moving the Heisenberg cut all the way up, until everything normally considered to be part of the material world built of atoms and molecules, and of the electromagnetic and gravitational fields that they generate, were placed below the cut and were described in quantum mechanical terms, whereas our conscious experiences, including our perceptions, were generally described in psychological terms, *but with our perceptions of the external world expressed in the usual way associated with the concepts of classical physics.*

The theory thus becomes a genuine psychophysical theory with the boundary between our conscious experiences and the underlying atom-based physical world lying at the mind-brain interface. A key aspect of the theory thus becomes a description of what is happening at the mind-brain interface between the experience-based mental aspects and the quantum mechanically described atom-based aspects of the evolving reality.

Classical Description, Oscillations, and the Quantum Mechanical "Coherent States" of the Electromagnetic Field

What we see, do, and intend to do is described at the mental level in classical terms, but at the brain level in quantum mechanical terms. This need to correlate a classical mental description to a naturally corresponding quantum counterpart at the

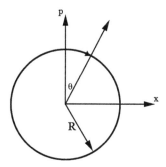

Fig. F1 SHO Kinematics

mind-brain interface is met by taking this connection to be via the well-known "coherent states" of quantum electrodynamics. These are quantum states that exhibit a simple harmonic oscillator (SHO) motion that is essentially identical to a classical SHO motion, except that the classical point particle is replaced by a minimum uncertainty Gaussian quantum wave packet whose center point follows the phase-space trajectory of the classical oscillating point (Fig. F1).

Our interest is in the possible influence upon the radius R of the mental choices made by the owner of the brain, within the framework of von Neumann's dynamical theory of the mind-brain connection.

Von Neumann's Dynamical Theory of the Mind-Brain Connection

The central problem in quantum mechanics is that the basic dynamical equation, the Schroedinger equation, generates not the actual evolving physical reality itself, but only a smear of potentialities for future actualities.

But then how does what actually occurs get picked out from the smear of potentialities?

"What becomes actual" is not picked out by nature acting alone. According to quantum mechanics, some subject/observer/agent must pose a question: "Is my immediately-to-appear experience Experience X?" Yes or No?. Nature immediately answers, and in the "Yes" case delivers Experience X to the observer. In either case, nature changes (instantaneously in a certain way) the entire physical world by eliminating all features that are incompatible with the delivered answer, Yes or No, that it has just chosen. This action takes care of the EPR correlations between outcomes of effectively-simultaneous far-apart experiments.

That choice of probing question on the part of some observer will single out some classically describable possibility. Quantum mathematics does not specify what question will be asked. The choice, according to quantum ideas, is "a free choice on the part of the observer", where 'free' emphasizes that the choice is not determined by the known laws of physics. The fact that what question is asked is classically describable accords with the idea that this choice comes from the mental realm of the observer.

The Pertinent Numbers

The measured general numbers for the Cortex are:

Size of computational unit: $Sz = (1/20 \times 1/20 \times 2.4) \times 10^{-9} m^3$

$$= 6 \times 10^{-12}$$

[Ref. Brain 125(5), 935–951, Buxheoven & Casanova.]

Strength of the magnetic field: H = ½ picotesla

SHO frequency: 20 Hz

R = Radius of SHO orbit in the usual Modified Phase Space in which the coordinate variable is

$y = $ [sq.root $(\bar{h}/m\omega)$] times coordinate variable X, (X meters) (m = 1 kg) (mks units) (angular velocity ω in radians per second) [20 Hz => $\omega = 20 \times 2pi$]

[Ref. Wikipedia: Quantum harmonic oscillator]

$$\text{Energy} = 2 \times \left(\tfrac{1}{2}H^2/\mu_0\right) \times Sz = \omega \times \bar{h} \times R^2$$

$$\mu_0 = 4\pi \times 10^{-7}\bar{h} = 10^{(-34)} \text{ in mks units}$$

$$\begin{aligned}
\text{Energy} &= \left(\tfrac{1}{4}10^{(-12)}\right)^2 \times (1/4\pi) \times 10^7 \times 6 \times 10^{-12} \\
&= \tfrac{1}{4}(6/4\pi) \times 10^{(-29)} \\
&= \tfrac{1}{4}(60/4\pi) \times 10^{-(30)} \\
&= 15/4\pi \times 10^{(-30)} \\
\sim &= 10 \times 10(-31)
\end{aligned}$$

$$\begin{aligned}
\text{Energy} &= \omega \times \bar{h} \times R^2 \\
&= 20 \times 2\pi \times \left(10^{-34}\right) \times R^2 \\
&= (1/8) \times \left(10^{-31}\right) \times R^2
\end{aligned}$$

$$R^2 = 80 \quad R \sim = 9$$

This number indicate that the process is at the quantum scale, and that a small change ΔR in R can give a significant change in the pertinent energy R^2.

The Quantum Zeno Mechanism for Mental Control of Bodily Action, and Recent Empirical Evidence from Neuroscience

Let $\Psi(R)$ be the quantum SHO state whose center is located at radius R on the rotating ray that represents the 20 Hz EM oscillation in the computational unit.

If the current state is Rho(R), and one asks the question "Is the state $\Psi(R + \Delta)$?", then the probability that the answer is "Yes" is $|<P(R)|\Psi(R+\Delta)>|^2$, which for small Δ is $(1 - \Delta^2)$.

If Δ is small, then the number N of probing questions that one can ask such that with 90% probability the answers will all be "Yes", so that the intended increase in R will occur with probability at least 90%, is therefore the N such that N $\Delta^2 = 1/10$, or $N = \left(1/(10\Delta^2)\right)$. Hence the agent can achieve an intended objective $\Delta R = N\,\Delta$ with 90% certainty if $\Delta R = 1/(10\,\Delta)$ for small Δ.

A pertinent question is: What rates of probing actions are needed in order to account, via this QZE mechanism, for the correlations found recently in neuroscience between intended actions and brain activity? Do we need extremely rapid probing rates?

Reference [1] describes statistically significant correlations between instructed manual motions of monkeys (which I am considering to be governed by QZE) and electromagnetic activity in the motor cortex. Figure 1c at 20 Hz and near 100 ms shows significant structure occurring over a 10 ms interval.

Ref. [1] Nature neuroscience, Propagating waves mediate information transfer in the motor cortex, Doug Rubino, Kay Robbins, & Nicholas Hatsopoulos. (Full text available on Wikipedia.)

If one wishes to achieve a certain increase ΔR over a ten millisecond interval with a uniform set of increases Δ this $\Delta = 1/(10\ \Delta R)$, and the number of steps needed is

$$N = (1/10)\Delta^{(-2)} = (1/10)(10\,\Delta R)^2 = 10 \times (\Delta R)^2$$

To achieve a unit change ΔR in R this number is 10, and the probing actions need occur only once each millisecond. These are normal time scales for neuroscience.

By comparing Fig. 4b, with field potentia 0.5 mV to Fig. 7 with field potentia 20 μV = 0.02 mV I surmise that the R that we are dealing with is probably much less than 9, and that ΔR is therefore not large. So the empirical numbers suggest that the results shown in Ref. [1] are probably concordant with the proposed understanding of von Neumann's theory of the mental causation of bodily action.

Appendix G
Transcending Newton's Legacy[2]

Science can influence our lives in many ways. The influence via technology is evident. But influence through effects on social institutions, such as church and government, can also be important. For example, the well-known influence of Newton's idea of "laws" upon the U.S. constitution could, in view of the immense influence of government upon our rights and freedoms, and upon our economic environment, be exerting tremendous influence upon our lives.

But more important than either of these is probably the influence of science upon our idea of what we are; upon our idea of our place in the universe, and of our connection to the power that forms it. For our aspirations and our values spring, in the end, from our idea of what we are, and nothing is as important in our lives as the nature of the ideas that motivate our actions, and the actions of others.

Science was transformed during the twentieth century by three revolutionary developments: the special theory of relativity, the general theory of relativity, and quantum theory. These developments altered not only scientific practice, but also our ideas about the nature of science and the nature of the world itself. I shall discuss here these three developments with regard to both their essential differences from classical Newtonian science, and their potential impact upon the human condition.

Newtonian Science

Newtonian science must be distinguished from the full thought of Isaac Newton. The former may be characterized by the following three conditions:

1. Absolute Time and Absolute Space. Newton's starting point is the idea of a "true" time and a "true" space. Each is independent of anything external to it,

[2]Invited talk given at the conference "Newton's Legacy: A Symposium on the Origins and Influence of Newtonian Science" Tulane University, November 12–14, 1987. This work, as LBL-24322, was supported by the Director, Office of Energy Research, Office of High Energy and Nuclear Physics, Division of High Energy Physics of the U.S. Department of Energy under Contract DE-AC03i6SF00098.

© Springer International Publishing AG 2017
H.P. Stapp, *Quantum Theory and Free Will*,
DOI 10.1007/978-3-319-58301-3

and has an inherent quality of uniformity or homogeneity. These two "absolutes" are contrasted by Newton to their "relative", or "apparent" counterparts, which we can grasp through our senses, and can measure by means of clocks and rulers.

2. Local Ontology. Absolute space is conceived by Newton to be populated with small bodies or particles that move with the passage of absolute time.
3. Fixed Laws of Motion. The motions of the particles are governed by "laws". These laws cause the locations and velocities of all particles at all times to be determined by the locations and positions of all particles at any single time. The world is therefore deterministic: its condition at one time determines it condition for all time.

These features of Newtonian science give us a picture of the universe called the Mechanical World-View. According to this view the universe consists of nothing but objectively existing particles moving through absolute space in the course of absolute time in a way completely determined by fixed laws of motion.

This picture of the world is mathematical: the objects are described mathematically, by numbers that give the locations and velocities of all the particles. Moreover, the laws that govern these numbers are mathematical. That Newton aspired to the creation of a mathematical picture of Nature is proclaimed by his title: Mathematical Principles of Natural Philosophy.

Three Problems

Some difficulties with this picture of Nature were evident from the start.
    I mention three:

(1) Action-at-a-distance
(2) Creation
(3) Freedom

(1) **The Problem of Action-at-a-Distance.**

The centerpiece of Newton's science is the law of gravity. According to this law, every body in the universe acts instantaneously upon every other one, even though they be separated by astronomical distances. Newton's recognition of a problem with this idea is expressed clearly in his famous assertion: "That one body can act upon another at a distance through the vacuum without the mediation of anything else ... is to me so great an absurdity that I believe no man, who has in philosophical matters a competent faculty of thinking, can ever fall into it". The ontology set forth in the Principia has, however, nothing to mediate the force of gravity. Newton worked hard to find carrier for gravity compatible with the available empirical evidence, much of which came from his own experiments. Finding in the end nothing that met his standards he declared: "hypothesis non fingo": I frame no hypothesis.

Two contrasting attitudes toward physical theory can thus be found in Newton's thinking. One attitude reflects his basic overriding commitment to search for truth about Nature. This commitment is massively displayed by his extensive researches into alchemical and theological questions pertaining to the constitution of Nature, by his choice of title mentioned above, and by his careful attention, in the formulation of his principles, to philosophical and ontological details. The second attitude goes with his "hypothesis non fingo". This declaration entails that his theory, as it stood, must, strictly speaking, be construed not as an ontological description of Nature itself, but merely as a codification of connections between measurements. The theory must be viewed as a system of rules that describes how our observations hang together, not as a description of the underlying reality.

These two contrasting attitudes toward physical theories will be the focal point of my discussion of how Newton's ideas fared in the twentieth century. The issue concerns two views of the nature of physical theory. One view holds that basic physical theory ought to provide a description of the real stuff from which the universe is constructed—it should describe the ultimate things-in-themselves.

The second view holds that physical theories should deal fundamentally with quantities that can be measured—they should merely codify the structural features of measurable phenomena.

(2)   The Problem of Creation.

The second problem is the problem of creation. Given the Newtonian precepts two questions immediately arise.

1. What fixed the nature of the particles and their laws of interaction?
2. What fixed the initial locations and velocities of the particles in the universe?

Within Newtonian science these two questions are insoluble. Thus from the perspective of the first attitude described above, which holds that basic physical theory should describe the real world, the account provided by Newtonian science is deficient, for it requires something external to the physical world it describes: it needs something to set up the system and fix the undetermined parameters.

From the second point of view, which is that science should merely codify, not explain, this problem of creation might seem to be no problem at all. But the problem is then with the point of view itself, which tends to close off the pursuit of the further knowledge. For, today, within the quantum theoretical framework, physicists are examining theories that purport to answer the first of the questions raised above, just on the basis of self-consistency. Moreover, the second question is moving into science in connection with studies pertaining to the birth of the universe—the big bang. The question is therefore this: "To what can science aspire?" Can it cope with the problem of creation, or must it remain forever mute on this basic question?

(3) The Problem of Freedom.

Beyond these questions is one far more pressing to man. The mechanistic world-view proclaimed by Newtonian science, and "validated" by its technological success, insists that all creative activity ceased with the birth of the universe. It tells us that we are now living in a "dead" universe that grinds inexorably along a path pre-ordained at the birth of the universe, and held in place by immutable laws of nature. Thus any notion that we can, by our efforts, act to bring into being one state of affairs rather than another is sheer illusion and fantasy.

This dreary view is proclaimed in the name of science, and is backed by its authority. Banished, together with freedom, is any rational notion of human responsibility. For responsibility can be placed only where freedom lies, and according to the precepts of Newtonian science all freedom expired when the universe was born.

I shall return to these questions from the perspective of twentieth century science. But first an essential stepping stone from the ideas of Newton to those of the twentieth century must be described.

(4) Galileo and Lorentz.

The laws of Newton have a simple consequence: given one possible universe, evolving in accordance with Newton's laws, it is possible to construct another in a simple way—just add to every particle in the universe any single common velocity. Then all separations between particles are left unchanged, and, according to Newton's laws, this shifted state of affairs will perpetuate itself through all time. This property is called Galilean invariance.

In 1873 James Clark Maxwell proposed a theory of electric and magnetic forces that was wonderfully beautiful and marvelously successful. This theory did for electricity and magnetism what Newton had tried to do for gravity: it explained the forces between charged particles in terms of changes that propagate from point to neighboring point, thus abolishing the need, in electricity and magnetism, for action-at-a-distance. However, the theory of Maxwell was characterized by a certain maximum speed, the velocity of light in vacuum. According to this theory no charged particle could move faster than this maximum speed. Consequently, the property of Galilean invariance was lost. However, Maxwell's theory had a substitute, which involved the characteristic maximum speed, the velocity of light. This new property, called Lorentz invariance, was to play a crucial role in what lay ahead.

(5) Absolute Versus Relative in Twentieth Century Science. The Special Theory of Relativity.

According to Newton's idea of absolute time one can assert that if A and B are two events, each of negligible duration, then either A is earlier than B, or B is earlier than A, or they are simultaneous. The truth of any such assertion, say that "A is earlier than B", is absolute: it does not depend upon anything else.

Consider, however, two such events A and B situated so that nothing can move from either event to the other without traveling faster than light. In this case one cannot determine by direct observation (say the observation of one event from the location of the other) which event occurs earlier than the other. One might expect that such a determination could be achieved by indirect means. However, Einstein showed that if all phenomena in Nature enjoyed the Lorentz invariance property mentioned above then it would be impossible in principle to determine from empirical data which of the two events occurred first.

The Lorentz invariance property seemed to hold universally (phenomena associated with gravity excepted, since Newton's theory of gravity needed to be reformulated along the lines of Maxwell's treatment of the electric force). Consequently, Newton's idea of absolute time seemed to bring into physical theory a property that in principle could have no correlate in observable phenomena. Einstein therefore proposed that physical theory be based not on absolute time and absolute space, as Newton had proposed, but rather upon a space-time structure defined by idealized readings of clocks and rulers.

The resulting theory is the special theory of relativity. Physicists quickly accepted this idea, which produced economy in notation and conception. Thus they replaced the absolutes of Newton by their relative counterparts.

(6) Quantum Theory.

Quantum theory is another twentieth century development that makes measurements primary. It carries the shift from absolute to relative even further than the special theory of relativity. For, according to the orthodox view of quantum theorists, not only must the underlying space-time framework be understood in terms of results of possible measurements, but, in fact, the entire mathematical formalism of quantum theory must be interpreted merely as a tool for making predictions about results of measurements. This view of quantum theory arose from its historical origin and its intrinsic form. But it is sustained by a reason far more compelling than mere "economy": every known ontology that is compatible with the phenomena, as codified by quantum theory, is "grotesque" in some way. Orthodox physicists, reluctant to embrace the grotesque, prefer to adopt a rational stance that separates the predictive mathematical formalism, and the associated scientific practices, from ontological speculations that lack empirical support.

(7) Conversation Between Einstein and Heisenberg.

1. Werner Heisenberg was the principal creator of the formalism of quantum theory. He has given an account of an interesting encounter with Einstein.

2. He prefaces this account with a brief description of the genesis of quantum theory: He, Heisenberg, reflecting upon Einstein's claim that a physical theory should contain only quantities that can be directly measured, and realizing that orbits of electrons inside atoms cannot be

observed, was led to discover rules that directly connect various measurable quantities pertaining to experiments performed on atomic systems, without ever referring to unobservable orbits.

Early in 1926 Heisenberg described this new quantum theory at a symposium in Berlin attended by Einstein. Later, in private, Einstein objected to the feature that the atomic orbits were left out. For, he argued, the trajectories of electrons in cloud chambers can be observed, so it seems absurd to allow them there but not inside atoms. Heisenberg, citing the nonobservability of orbits inside atoms, pointed out that he was merely following the philosophy that Einstein himself had used. To this Einstein replied: "Perhaps I did use such a philosophy earlier, and even wrote it, but it is nonsense all the same." Heisenberg was "astonished": Einstein had reversed himself on the idea with which he had revolutionized physics!

To find the probable cause of this "astonishing" reversal it is necessary only to look at what Einstein had done between the 1905 creation of special relativity and the 1925 creation of quantum theory. The special theory holds, as mentioned earlier, only to the extent that the effects of gravity can be ignored. It was necessary to generalize the special theory to the general case by incorporating a reformulation of Newton's theory of gravity along the lines of Maxwell's theory of the electric force.

Einstein undertook this task and in 1915 announced his general theory of relativity. Though this theory was a generalization of the special theory in many ways, it was fundamentally different. The focus was no longer on observers and results of measurements. The theory was about a space-time structure that exists by itself, governed by its own nature, without relation to anything external. It was about an "absolute" space-time structure. Einstein was driven during his ten-year search for the general theory not by an effort to codify data. He was driven by demands for rational coherence and by a general principle of equivalence. He sent his work to Born saying that no argument in favor of the theory would be given, since once the theory was understood no such argument would be needed.

Einstein had in this work gone beyond the need for "hypothesis non fingo". He had succeeded in doing what Newton had failed to do. He had discovered a mathematical description of something that could be regarded as Nature itself. The difficulty that defeated Newton, namely the action of gravity at a distance without any carrier, he had resolved by first combining: "Newton's absolute time and absolute space into an absolute space-time".

(8) Relaxing Newton's demand for uniformity, and finally imposing his mathematical laws in the form of conditions on deviations from uniformity: the presence of matter was represented by departures from uniformity—by distortions of space-time itself.

An important difference between Einstein's theory and that of Newton is that in Newton's theory time and space are independent of each other, and both are independent of matter. This creates, at least in principle, the possibility of space with nothing in it: an empty arena.

The idea of empty space has puzzled philosophers since antiquity: how can anything be nothing; that is a contradiction in terms. Thus Newton's predecessor Descartes takes extension, hence space, to be something that cannot exist without matter. Newton's contemporary Leibniz takes space to be merely a system of relations. Still, it remains puzzling that so much of the universe can be (almost) empty space if empty space is nothing at all.

Einstein's ontology gives a marvelous solution to this ancient puzzle. Instead of three intrinsically different things—time, space, and matter—whose connection must then, from a logical point of view, be ad hoc, hence puzzling, we have only one thing: inhomogeneous space-time. Considering the direction and achievements of Einstein's general theory of relativity one cannot be surprised that its creator should regard the philosophy of the creator of the special theory of relativity as "nonsense all the same".

The fate in the twentieth century of Newton's two absolutes is then this: the special theory of relativity replaced them by their relative counterparts, but the general theory resurrected them in a combined form that incorporates also the third element of Newton's ontology, matter. However, quantum theory represents a swing from the absolute back to the relative. For, according to the orthodox view, quantum theory must be viewed as codification of connections between measurable, or relative, quantities.

With this background in place, I turn now to the question of the impact of twentieth century science upon our ideas about Nature, and upon our ideas about ourselves.

(9) Impact of Quantum Theory Upon the Mechanistic World-View of Newtonian Science.

Quantum theory gives in general only statistical predictions. The question thus arises: Does Nature itself have genuinely stochastic or random elements? Bohr stated the orthodox position: We find, in practice, that even when we prepare an atomic system to the limits of our capabilities there is still a scatter in the results of certain experiments. Quantum theory gives predictions with a matching irreducible scatter. Thus the statistical character of the theory matches the statistical character of the facts. To say more than this is empirically unsupported speculation: quantum theory says nothing about determinism in Nature.

Quantum theory successfully describes and predicts phenomena on the basis of a mathematical description of atoms. Can we conclude that the world is built of atoms?

If one looks at the mathematical representation of these atoms one finds entities that must, according to the orthodox view, be interpreted only as parts of a computation of expectations pertaining to results of measurements. Thus the ontological foundation is shifted from the level of the atoms to the level of the devices that record these results, or perhaps even to the level of the observers who use these results to make computations. But the devices and observers are assumed to be built from atoms. So the ontological basis swings back to the atoms, etc.

These examples illustrate the difficulty in trying to draw ontological conclusions from a theory that must be interpreted merely as a tool for making predictions about connections between measurements.

(10)   Quantum Theory and Reality.

It is clear to everyone that we cannot pass with certainty from knowledge about the structure of phenomena to knowledge about the structure of the underlying reality. Accordingly, the orthodox interpretation of quantum theory tries to isolate, insofar as possible, the mathematical formalism, and the scientific practices associated with it, from more speculative activities: it tries to separate "science" from "natural philosophy". Science is concerned with measurable quantities, and with theoretical structures that codify the observable and testable connections between them. Natural philosophy concerns the conclusions that might reasonably be drawn about the form of the underlying reality on the basis of the evidence provided by science. The fact that Bohr and Heisenberg adhered to the view that the mathematical formalism of quantum theory should be viewed, strictly speaking, merely as a tool for making predictions pertaining to results of measurements in no way implies that they had no interest in the implications that quantum theory has in the realm of natural philosophy. In fact, each in his own way tried to draw from the data provided by quantum theory insights into the nature of the world that lies behind the phenomena.

(11)   Heisenberg's Ontology.

Heisenberg in his book "Physics and Philosophy" in the chapter on the Copenhagen interpretation actually sets forth an ontology. He begins with the words "If we want to describe what happens in an atomic event ...". He then goes on to describe an ontology in which the actual world is formed by "actual events", which occur only at the level of the macroscopic devices. But the objective world contains also something else. It contains "objective potentia". These "objective potentia" are objective tendencies for the actual events to occur. They are associated with the mathematical probabilities that occur in quantum theory. This ontological substructure gives nothing testable. So it is not "science".

But it gives us an informal way of "understanding" quantum theory. It gives us an idea of what is actually going on.

This ontology described by Heisenberg is not the only ontology compatible with the predictions of quantum theory. But it can be said to be the "most orthodox" ontology. Most quantum physicists probably think about quantum. phenomena informally in these terms: the quantum probability functions corresponds somehow to the **tendency** for the detector to register a particle, or the **tendency** for a grain in a photographic plate to register the absorption of a photon. The actual things occur only at the macroscopic level.

Heisenberg's ontology cannot be deduced from the phenomena, and is therefore speculative, and to be distinguished from science. However, I do not think it unreasonable to consider it seriously. All creation is certainly not simply a collection of measurements floating on nothing else, even though measurements are of particular interest in science, and are the source of our most precise information about the world.

The reason it is interesting to consider the ontologies suggested by the structure of phenomena as codified by quantum theory, and compatible with that structure, is that the conditions thus imposed on ontologies are so restrictive: there is no known ontology that is compatible with the conditions on phenomena imposed by quantum theory that is not "grotesque" in the minds of conservative thinkers. This means that quantum theory has shown us that the world is not at all like what we had previously imagined it to be. It is not at all like the idea of it set forth in the mechanical world-view, formerly (pre 1900) promulgated in the name of science, and still largely dominating the prevailing idea of what science tells us. So any curious person must naturally be led to ask: What idea of the world is compatible with the data provided by science?

(12)   World-View Arising From Heisenberg's Ontology

Heisenberg's ontology is the most-orthodox, and, in my opinion, the most reasonable, of the known ontologies that are compatible with the predictions of quantum theory. In the remainder of this article I shall describe the principal features of the picture of Nature that arises from this quantum ontology.

1. The World is Nonlocal.

   Macroscopically separated parts of the universe are linked together in a way that involves strong faster-than-light connections that do not fall off with increasing spatial separation. This nonlocal aspect is the "grotesque" feature of this ontology that makes it unacceptable to conservative thought.

2. Creation is Distributed Over All Time.

   In the quantum ontology the objective potentia are represented by the quantum probability function. At each stage the quantum potentia give tendencies for the next actual event. The occurrence of this next actual event is represented by a "collapse" of the potentia to a new form. The interplay of the Heisenberg uncertainty relations and the Heisenberg equations of motion is such that, even though each successive event effectively closes off certain possibilities, by making fixed and settled things that had formerly been unfixed, still, each event creates new potentialities and possibilities. Consequently, the process of fixing the unspecified degrees of freedom, which in classical physics occurs all at once, at the creation of the universe, is, in the quantum ontology, by virtue of its mathematical structure, a process that can never close off the possibility of its further action. Thus in the quantum ontology, the creative process, in which things formerly unfixed become fixed and settled, does not expire at the birth of the universe, but extends rather over all time.

3. Two Kinds of Time.

The quantum ontology has two different times. The first is Einstein Time, which joins with space to form Einstein's space-time. The second is Process Time. I shall now explain the difference. The "numbers" that appeared in Newton's theory, and which described the positions and velocities of the particles, are replaced in quantum theory by

(13)   "operators", which evolve in accordance with equations, called Heisenberg's equations of motion.

The evolution of the quantum operators in accordance with Heisenberg's equations of motion is evolution in Einstein time. This evolution generates an association of operators with space-time points: every space-time point, from the infinite past to the infinite future, is associated with a fixed set of operators.

The space-time structure just described is a structure of quantum operators. To obtain the potentia one must take these operators in conjunction with something called the Heisenberg state vector. The Heisenberg state vector does not depend on space-time: it refers to all of space-time. But it combines with the operators associated with any space-time point to produce numerical potentia associated with that space-time point.

Each actual event is associated with a "quantum jump" of the Heisenberg state vector. Thus each actual event induces a sudden jump in the potentia. This jump occurs at every space-time point. The sequence of quantum jumps defines a time that is different from Einstein time. It is called Process Time. Evolution in process time generates change or evolution of the "actual", whereas evolution in Einstein time generates the evolution of the "potentia". Thus the deterministic laws of evolution are not binding on our future, for they determine the evolution of the potentialities, not the actual events themselves.

(14)   Meaning in the Quantum Universe.

The creative process is represented in the quantum ontology by the sequence of jumps in the quantum potentia. These potentia are objective tendencies, which tend to make the statistical predictions of quantum theory hold under appropriate conditions. But the question arises: What determines the actual course of events? That is, what determines, in a given actual instance, whether things will be fixed in one way or another? Heisenberg's ontology leaves that crucial question unanswered. Hence the ontology, as presently understood, is incomplete.

At first', it might seem that, in any case, the choice of what actually happens is either deterministically fixed by what has gone before, or has an element of true randomness or wildness. In either case, the ontology would appear to provide no possibility for a meaningful universe: either we would have simply a new determinism, which would render the universe just as "dead", and devoid of possible meaning, as the world of Newtonian mechanics, or there would be an element of randomness, which could hardly add meaning. Thus we are apparently still trapped between the two horns, determinism or randomness, of the usual dilemma of the impossibility of a meaningful universe.

To have meaning a choice must have intentionality: it must exist in conjunction with an image of the future that it acts to block or help bring into being. Any choice that does not refer in this way to the future is a meaningless choice.

In the Newtonian picture the future does not exist in the present, and hence it cannot enter into a present event or choice. Moreover, the future cannot be changed by any event or choice.

But in the quantum ontology the future does exist objectively in the actual present, albeit as potentia. Thus the future can enter into the present event. This event can by altering the potentia for the future events, effectively block or help bring into being a chosen state of affairs. In this sense a quantum event can have effective intentionality and meaning.

(15) Man in the Quantum Universe.

The role of man in the universe is tied to the mind-body problem. From the perspective of the quantum ontology the brain is a macroscopic system similar to a measuring device. The function of the brain is to organize input, and then make a decision that initiates an appropriate action. According to the brain-device analogy this decision is represented as a quantum jump.

Just as in the case of a measuring device, this quantum jump is a macroscopic event: the whole brain, or some macroscopic part of it, is involved.

The problem of understanding, within the framework provided by classical physics, the connection between consciousness and the physics of the brain has been described in some quotations cited by William James: "The passage from the physics of the brain to the corresponding facts of consciousness is unthinkable. Granted that a definite thought and a definite molecular action in the brain occur simultaneously; we do not possess the intellectual organ, nor apparently any rudiment of the organ, which would enable us to pass, by a process of reasoning, from one to the other." (Tyndall). Or

Suppose it to have become quite clear that a shock in consciousness and a molecular action are the subjective and objective faces of the same thing; we continue utterly incapable of uniting the two, so as to conceive that reality of which they are the opposite faces (Spencer).

The quantum ontology does have an analog of the classical motions of molecules moving in accordance with Newton's laws: it is the evolution of the corresponding quantum operators in accordance with Heisenberg's equations. However, the quantum ontology has, also, something else, which has no counterpart or analog in classical physics: the actual event. Within the quantum ontology the conscious event and the physical event can be naturally understood as the psychological and physical faces of the same event, namely the event of selecting and initiating a course of action. On the psychological side there is the felt or conscious event of selecting and initiating this action, and on the physical side there is the physical collapse of the potentia, which selects and initiates this action: the physical brain, as represented in quantum mechanics, collapses to a state in which the instructions that initiate the particular course of action are actualized.

The connection between these two events is not an ad hoc and arbitrary identification of things as totally disparate as, on the one hand, a motion of billions of separate molecules, and, on the other hand, a unified conscious act.

(16) It is, rather, the association and identification of the felt event with the physical event that represents, within the quantum ontology, exactly the change that is felt. In this way conscious events become special instances of the actual events that, according to the quantum ontology, form the fabric of the entire actual universe.

Conclusion

Quantum theory had several founders who had different opinions regarding ontology. Hence it may not entail any specific ontology. However, the most reasonable and well-defined of the prominent quantum ontologies is, in my opinion, the one that combines the claim of Heisenberg that we are dealing with a choice on the part of the observer pertaining to which experiment will be performed, and the claim of Dirac that we are dealing with a choice on the part of "nature" pertaining to which response will be delivered, as formalized by von Neumann's theory of measurement, dubbed "orthodox" by Eugene Wigner. and in which the evolving density matrix represents "potential", expressed as probability for specified experienced response.

The chief features of the world that flow from this ontology are:

1. It is nonlocal: there is some sort of nonseparability of spatially separated parts of the universe.
2. It is creative: the fixing of previously unsettled matters is a continuing process; creativity did not expire with the birth of the universe.
3. It could be complete: no aspect of reality not represented within the quantum ontology seems necessarily required.
4. It allows meaning: choices can have intentionality, hence meaning.

In everyone of these essential aspects the world-view provided by the quantum ontology is the reverse of the one provided by pre-twentieth century science. Consequently, modern science provides man with a vision of himself that is altogether different from, and far more inspirational and philosophically fertile than the one proclaimed in the name of Newtonian science.

No longer is man reduced to a cog in a giant machine, an impotent witness to a pre-ordained fate in some senseless charade. Rather, he appears, most naturally, within the framework of present-day science, as an aspect of a fundamentally nonseparable universe that is creation itself, both as noun and verb, a creative process that unites in an intelligible way the mental and physical aspects of Nature, and is moreover endowed in principle with the capacity to suffuse its evolving form with meaning.

References

1. von Neumann J (1932) Mathematische grundlagen der quantenmechanik. Springer, Heidelberg (Translated as Mathematical foundations of quantum mechanics, Princeton University Press, Princeton NJ, 1955. Chap. IV)
2. Crick F, Koch C (2002) The problem of consciousness. In: Scientific American (Special Issue: The Hidden Mind)
3. Damasio A (2002) How the brain creates the mind. In: Scientific American (Special Issue: The Hidden Mind)
4. Einstein A (1949), Schilpp PA (ed) Albert Einstein: philosopher-scientist. Tudor, New York, p 674
5. Einstein A (1949), Schilpp PA (ed) Albert Einstein: philosopher-scientist. Tudor, New York, p 670
6. Einstein A, Podolsky B, Rosen N (1935) Can quantum mechanical description of physical reality be considered complete? Phys Rev 47:777–780
7. Bohr N (1935) Can quantum mechanical description of physical reality be considered complete? Phys Rev 48:696–702
8. Bell JS (1964) On the Einstein-Podolsky-Rosen paradox. Physics 1:195–200
9. Wheeler JA (1987) The 'past' and the delayed-choice double-slit experiment. In: Marlow AR (ed) Mathematical foundations of quantum theory. Academic Press, New York, pp 9–48
10. Kim YH, Yu R, Kulik SP, Shih Y, Scully MO (2000) Delayed "choice" quantum eraser. Phys Rev Lett 84:1–5
11. Bohm D (1952) A suggested interpretation of quantum theory in terms of hidden variables. Phys Rev 85:166–179
12. Hawking S, Mlodinow L (2010) The grand design. Bantam, New York, p 140
13. Griffiths R (2002) Consistent quantum theory. Cambridge University Press, Cambridge, U.K.
14. Stapp HP (2012) Quantum locality? Found Phys. doi:10.1007/s10701-012-9632-1. arXiv. org/abs/arXiv:1111.5364
15. Stapp HP (2007, 2011) Mindful universe: quantum mechanics and the participating observer. Springer, Berlin, Heidelberg (Appendix G)
16. Stapp HP (1993, 2003, 2009) Mind, matter, and quantum mechanics. Springer, Berlin, Heidelberg
17. Stapp IIP (2001) Quantum theory and the role of mind in nature. Found Phys 31:1465–1499. arxiv.org/abs/quant-ph/0103043
18. Stapp HP (1977) Are superluminal connections necessary? Nuovo Cimento 40B:191–205
19. Stapp HP, Philosopy of mind and the problem of freewill in the light of quantum mechanics. arxiv.org/abs/0805.0116
20. James W (1950) The principles of psychology, vol 1. Dover, New York, p 136 (Original Henry Holt, 1890)
21. Schurger A, Sitt JD, Dehaene S, An accumulator model for self-initiated neuralactivity prior to self-initiated movement (PNAS-2012-Schurger-1210467109)

© Springer International Publishing AG 2017
H.P. Stapp, *Quantum Theory and Free Will*,
DOI 10.1007/978-3-319-58301-3

22. Einstein A (1949), Schilpp PA (ed) Albert Einstein: philosopher-scientist. Tudor, New York, p 85
23. Bell J (1987) Speakable and unspeakable in quantum mechanics. Cambridge University Press, Cambridge, UK, p 1, 15
24. Bem DJ (2011) Feeling the future: experimental evidence for anomalous retroactive influences on cognition and affect. J Personal Soc Psychol 100:407–425
25. New York Times. http://www.nytimes.com/roomfordebate/2011/01/06/the-esp-study-when-science-goes-psychic/a-cutoff-for-craziness
26. von Neumann J (1955) Mathematical foundations of quantum theory. Princeton University Press, Princeton
27. Heisenberg W (1958) The representation of nature in contemporary physics. Daedalus 87:95–108
28. Bohr N (1963) Essays 1958/1962 on atomic physics and human knowledge. Wiley, New York, p 60
29. Heisenberg W, in Appendix A of ref. [30]
30. Stapp HP (1972) The copenhagen interpretation. Am J Phys 40:1098–1116 (Reprinted as Chap. 3 in Mind, Matter, and Quantum Mechanics. Springer, Berlin, Heidelberg, New York, 1993, 2003, 2009)
31. Einstein A (1951), Schilpp PA (ed) Albert Einstein: philosopher-scientist. Tudor, New York, p 670
32. Wheeler JA (1987) The 'past' and the delayed-choice double-slit experiment. In: Marlow AR (ed) Mathematical foundations of quantum theory. Academic Press, New York, p 9–48
33. Kim YH, Yu R, Kulik SP, Shih Y, Scully MO (2000) Delayed "choice" quantum eraser. Phys Rev Lett 84:1–5
34. Hawking S, Mlodinow L (2010) The grand design. Bantam, New York, p 140
35. Stapp HP (1986) Process time and einstein time. In: Griffiths DR (ed) Physics and the ultimate significance of time. SUNY Press, Albany
36. Tomonaga S (1946) On a relativistically invariant formulation of the quantum theory of wave fields. Prog Theor Phys 1:27–42
37. Schwinger J (1951) The theory of quantized fields I. Phys Rev 82:914–927
38. Stapp HP (2007, 2011) Mindful universe: quantum mechanics and the participating observer. Springer, Berlin, Heidelberg (Chap. 13)
39. Stapp HP (2001) Quantum theory and the role of mind in nature. Found Phys 31:1465–1499 [arxiv.org/abs/quant-ph/0103043]
40. Newton I (1691) A letter to Richard Bentley
41. Heisenberg W (1983) Traditions in science. Seabury Press, New York
42. Heisenberg W (1958) Physics and philosophy. Harper & Row, New York (Chap. III. See also, David Bohm (1951) Quantum theory. Prentice-Hall, New York, Chap. 8)
43. James W (1950) The principles of psychology, vol 1. Dover, New York, p 147

Lightning Source UK Ltd.
Milton Keynes UK
UKHW021434250820
368772UK00003B/27